Whence and Wherefore

Also by Sir Bernard Lovell:

Science and Civilisation
World Power Resources and Social Development
Radio Astronomy
Meteor Astronomy
The Exploration of Space by Radio
The Individual and the Universe (BBC Reith Lectures)
The Exploration of Outer Space (Gregynog Lectures)
Discovering the Universe
Our Present Knowledge of the Universe
The Explosion of Science: The Physical Universe (ed. with
    T. Margerison)
The Story of Jodrell Bank
*The Origins and International Economics of Space
    Exploration*
Out of the Zenith
Man's Relation to the Universe
P.M.S. Blackett: A Biographical Memoir

Also by Zev Zahavy:

Independence and Modernity
The Lesser Assembly
A Space Age Guide for the Perplexed
Atheist Heresy and the Utopian Vision
Chaplain on Wings

JACKET PHOTO (Courtesy of NASA): A pro-
foundly remarkable view of the Earth's sphere,
photographed from the Apollo 17 spacecraft, depicts
the geographic area from the Mediterranean Sea
and the Arabian Peninsula down the entire coast-
line of Africa to the Antarctica south polar ice cap.
The Asian mainland is visible on the horizon
toward the northeast. A heavy cloud cover extends
along the Southern Hemisphere.

Whence and Wherefore

The Cosmological Destiny of Man
Scientifically and Philosophically
Considered

by Zev Zahavy

Comprising an analysis relating to the significant essay
"In the Centre of Immensities"
by the distinguished Professor
Sir Bernard Lovell
University of Manchester, England

Sage Books
New York

© 1978 by Zev Zahavy

First Printing
A.S. Barnes and Co., Inc.
1978

Second Printing
Sage Books
1982

Library of Congress Cataloging in Publication Data
Zahavy, Zev.
Whence and wherefore.

"In the centre of immensities, by Sir Bernard Lovell": p.
Includes index.
1. Cosmology. 2. Man. 3. Science—Philosophy.
4. Religion and science—1946–   I. Lovell, Alfred Charles
Bernard, Sir. 1913–   In the center of immensities. 1978.
II. Title.
BD511.Z3 1978   113   77-84786
ISBN 0-498-02190-4

To
The Great and Gracious
Master of the Universe
Who made it all possible

About the Author

Professor Zev Zahavy, Ph.D., has written extensively on philosophical topics, and he has engaged in considerable research in the area of cosmology. He has studied on the graduate level with leading American and British figures in the field of philosophy and religion. His variegated courses in the City University of New York have ranged from a presentation of classical intellectual literature to an analysis of modern existentialism.

Contents

Introduction

Now that man is a bona fide space traveler, it appears to be a propitious time to call for a new chapter in the enduring search for ultimate truths.

The technological record of human progress from the ancient period of a Sumerian bronze age to the current era of complex, multinational, computerized societies is a magnificent tribute to the impressive ingenuity of the human intellect. Man may justifiably take pride in his vast array of technological triumphs.

Certainly, the human specimen has traversed a long, rugged road since the early days of the ox-drawn plow and that prosaic lethargic era when simple mathematics were inscribed in fundamental ideograms upon cuneiform tablets. Today, man spans oceans and continents in a few short hours; communicates with the speed of light; harnesses the nuclear might of the unseen atom; prepares to visit neighboring planets; and charts the course of stellar phenomena at distant outposts of his universe.

Nevertheless, there is something radically amiss. Man's vainglory is tempered by the sordid fact that his society teeters perilously on the brink of perdition and ruination.

It is a pathetic situation. Is this a logical reward for

more than five thousand years of patient searching and suffering? Must civilization emerge as nothing more than a cruel, self-inflicted hoax?

Of what value is the stunning record of technological triumphs if, consequently, the human soul languishes in a more harrowing nebulosity of existentialist despair? Why must the psyche of twentieth-century man endure the pangs of depressing anguish? Why does it seem that many more individuals are finding life futile and meaningless?

Judging from a general assessment of the present, and endeavoring a simple prognosis for the future, man's experiment with a radically secular society has proved to be a dismal failure. The secular life has spawned a brutally amoral milieu. Human freedom has been translated into a cunning device for fiendish dominion and ruthless guile. One is tempted to conclude that species *Homo sapiens* is fast reverting back to the degrading rank of a savage beast.

Whereas God was once the chief inspiration for decent living, today brazen souls snidely allude to the Divine One's demise. There is a diabolical ulterior motive behind the attempt to exclude God from appearing in the human picture. By publishing God's obituary, man hopes to emerge as the uncontested supreme entity in the universe. With God out of the way, man believes that he is free to pamper his every conceit and caprice.

But how long can man hide from God, and how successfully can man fare in the parole of his own recognizance? Can man prod his own Freudian superego to formulate the necessary humanist recipe for a safe and sane society if all he has to fall back on is his ferocious id? Can man truly hope to guarantee his survival in a civilization that declares that God is obsolete? In the contemporary secularist environment that man has fashioned, and that predominantly ignores the divine presence, is man more secure —or is he more vulnerable to self-extinction?

12

We are living at a time when the existentialist plight of man is at a critical stage. The arsenals are overflowing with sophisticated weaponry that could very well wipe civilization from the face of the globe. The possibility of wholesale decimation is a real threat to the coming generations. Yet, the truth of the matter is that man does not want to be expunged from his little home in the solar system. Can man deter the ominous threat of an Armageddon?

If ever there was a time for serious introspection, it is now. If ever there was a time for a healing of breaches and a binding-up of fratricidal wounds, it is now. If ever there was a time for people of differing nationalities and cultures to extend a handclasp of sincere amity and harmony, it is now.

However, lest we succumb to the lure of futile idyllic dreams and utopian visions, let us consider a more realistic appraisal of the situation at hand. The scoundrels who presently abound in this world of ours are not going to relinquish their positions of power merely for the sake of serving the cause of world tranquility. Millions of innocents sadly must look forward to continued periods of inequity, hardship, and suffering. The present perpetrators of devilish designs will be succeeded by newer breeds of their ilk, and they will continue to impose upon the innocent multitude the venomous pattern of malicious hatred and falsehood, along with its practice of arbitrary discrimination and evil persecution.

Is it futile, then, to hope that the masses of people of good conscience everywhere may learn to pool their talents in the interest of human survival? If the common good is thwarted because of narrowly divisive national lines, perhaps in a world fast becoming increasingly interdependent, an endeavor should be made at least to develop a common camaraderie along humanistic lines.

13

Why should not the scientist and the theist walk side by side in service of an existentially impoverished humanity? After all, both are simultaneously pledged to the task of aiding mankind. Their partnership in pursuit of this humanitarian ideal will afford a much needed ray of hope for the future of civilization. The scientist and the theist do tread the same path toward a search for ultimate truth, although they operate on different levels of investigation.

It is in the hope that a new harmony may prevail between science and theology that prompted the writing of this book. There is reason to believe that modern science and modern theistic existentialism may develop into the leading cooperative components of a revitalized contemporary civilization. Assuming that knowledge is power, the combined might of both these disciplines should be enormous.

The raging battle for the control of man's mind is the most crucial struggle in our contemporary era. There is a lesson to be learned from the fastidious sponsors of evil. They resolutely exploit the full force of perverse propaganda with all the power at their command. Therefore, we must recognize the need to speak out, and we dare not remain silent. Unfortunately, however, the theistic existentialist position has been woefully inexpressive, and little has been done to counteract the specious sophistry of current insidious spokesmen. It is sad, because although the pronouncements of the purveyors of evil are widespread, their ulterior motive is narrowly egocentric and self-serving. Surely, the time is at hand when the clear voice of sincere seekers of truth must resound throughout the world unto all the inhabitants thereof.

Given the factual elucidations of science in harmonious coordination with profound theistic concepts, a newfound partnership may develop that will have no rival in the search for ultimate truth. More than ever, modern man thirsts for the wellspring of truth. Science and theology share alike an earnest dedication to the principles of truth and knowledge. Together, they may reawaken the dormant spirit of man and infuse his newly inspired soul with a sense of valid purpose. Life may gain a fresh impetus to reflect valid meaning for the human situation.

In line with such a design, this modest work proceeds to offer an analysis of the cosmic drama in two sections. Part one unveils the first act, and the curtain rises to present an enlightened diagnosis of the cosmological problem. This analysis comes from the pen of the world-renowned scientist and distinguished scholar, Sir Bernard Lovell, who has gained wide respect as Professor of Radio Astronomy and Director of the Experimental Station, Nuffield Radio Laboratories, Jodrell Bank, under the auspices of the University of Manchester in Cheshire, England. The essay, "In the Centre of Immensities," was first presented by Sir Bernard Lovell in August 1975, in Guildford Cathedral as his Presidential Address to the British Association, and it was reprinted in a shortened form in the *New York Times Magazine* on November 16, 1975, under the title "Whence."

The British Association for the Advancement of Science was organized in 1831 to give a more systematic direction to scientific inquiry, and "to promote the intercourse of the cultivators of science with one another and with foreign philosophers." Accordingly, "Wherefore," which forms the second part of this work, indulges in a theistic exposition reflecting modern existentialist principles, and since it is authored by an American professor of the City University of New York, it should qualify, from the British Association's point of view, as the product of a "foreign philosopher."

A note of deep gratitude is extended to the esteemed and respected scholar Sir Bernard Lovell, for granting permission to publish his superlative essay, which inclines toward the view that the essence of modern man derives from his primeval source in stellar space. The exploration of this theme from a scientific direction is accomplished with thoughtful insight and consummate skill.

As we proceed to the task of further enucleation, and after completing a brief journey through the portals of current scientific observation and judgment as outlined in the first section, we arrive at the second division of our work, which examines the grand cosmic drama in the domain of "Wherefore" and beyond.

It is felt in the "Wherefore" analysis that a consistently logical scientist should be able to harmonize theistic doctrines and scientific facts with dispassionate equanimity; and even as the new wave of space-age science progresses, it should not be unsettling to witness the blossoming of a hardy strain of resolute theistic scientists who may handily succeed the egregious atheistic breed of the twentieth century.

On the basis of this assumption, we may anticipate the time when the fad of atheism will commence to wane, and in its stead, the universal acknowledgement of God will become a *sine qua non* for future generations. It matters little what man concludes in eras of illogical fancies. God remains ever invincible, and He must inevitably emerge as the total victor. This concept, certainly, is hardly new, but its reaffirmation in the light of current atheistic trends that have produced a marked convolution of existentialist and psychological ego-deification, makes such a project an essential task for our age.

Long ago, wise King Solomon asserted that there is nothing new under the sun. It should not surprise us, therefore, to discern that the essence of the plot of the unfolding cosmic spectacle is quite succinctly reflected in the simple declaration of a dedicated prophet, whose words from the ancient past re-echo with urgent persistency today: "And God shall be sovereign over all the earth; in that day, God shall be one, and His name one." (Zechariah 14:9)

16

*Whence**

We are what we know about where we came from

In The Centre of Immensities*

by Sir Bernard Lovell, OBE, FRS

Presidential Address to the 137th Annual Meeting of the British Association for the Advancement of Science. Delivered in Guildford Cathedral, August 1975.

MAN'S ETERNAL QUEST

It is now nearly a century and a half since the British Association held its first meeting. The title of my address is taken from Thomas Carlyle's masterpiece written at that time. In *Sartor Resartus* he enquired, What is Man? who "sees and fashions for himself a Universe, with starry spaces and long thousands of years, . . . as it were, swathed-in and inextricably over-shrouded; yet it is sky-woven and worthy of a God. Stands he not thereby in the centre of Immensities, in the conflux of Eternities?"

One hundred and fifty years is a minute interval in the history of the human race. It is insignificant compared with the 4,500 million year history of the Earth—and the Earth itself is probably only half the age of the present universe. But my thesis is that in this short and cosmically insignificant time-span we have been brought to the centre of immensities—and that we stand there today not only in the world of our material existence, but also in the sphere of our intellectual endeavour to understand the universe and man's relation to it. It is to these two strangely contrasting, but vitally interwoven topics that I direct my thoughts this evening. The questions are whether man has reached a fundamental barrier in his attempt to comprehend the universe in a physical sense, and furthermore whether he can survive for long the consequences of the probing of the scientist to break through this barrier.

I can give you immediately a stark illustration of my concern. The subject known as exobiology is of great interest and importance. It relates to the problem of whether our life on Earth is unique in the whole universe. For a long time this topic has belonged to the realms of fiction. But in recent years, astronomers using radio telescopes have been studying the clouds of gas, known as nebulae, in our Milky Way. In the neighbourhood of hot stars these become luminous—the horsehead nebula in Orion is an example. They are not far away from the Sun by cosmical standards —only a few thousand light years, in the spiral arms of our own galaxy. We have understood for a considerable time that these nebulae are largely composed of hydrogen— probably remnant from the primeval material of the universe. Twelve years ago, to our great surprise, American radio astronomers discovered that the hydroxyl radical, OH, the combination of hydrogen and oxygen, was also present in these clouds of gas. Many different kinds of molecules have since been discovered in these nebulae—

20

including water. It is an amazing thought that if we could tune our eyes to a wavelength of 1.35 centimeters, we should find the emissions from these water molecules to be the most prominent objects in the sky. The complexity of other molecules in these gas clouds suggests, very forcibly, that the basic material for organic evolution exists in space.

But that is only one part of the story. More than 40 years ago when I was a young man attending my first British Association meeting in Aberdeen, Sir James Jeans in his presidential address gave a vivid and convincing picture of the formation of the solar system. He believed that the Sun was once an ordinary star without planets. Then a passing star came close enough to the Sun to drag out from it by gravitational attraction a great filament of gas which broke up and condensed into the planets. But within a year of Jeans' address, the American astronomer H N Russell, undermined all such encounter theories by pointing out that the star and the Sun would have to approach one another so closely that any planets resulting from the encounter would move in orbits thousands of times closer to the Sun than those we observe.

Another half century of observation and computation has led to the general view that the nebular hypothesis, proposed by Laplace 200 years ago, was at least partially correct. In his theory the Sun and planets condensed from a slowly rotating gaseous nebula. Now we think that processes involving the collision of dust and gas particles in the nebula, rather than condensation, lead to the formation of planets. It seems that in a few thousand million years a nebula surrounding a star could be transformed to a planetary system in this way. Naturally, one would expect violent collisions during the evolutionary process as evidenced by the craters on the Moon, and, as we have seen recently from the space probe photographs of the planets Mars and Mercury.

21

THE BIRTH OF A STAR

Our picture is then, that the planets began to accrete in a nebula of gas and dust surrounding the Sun about five thousand million years ago. How did the Sun acquire this nebula? In recent years, photographs have been obtained which are believed to show the early stages of stars forming in gas clouds in the spiral arms of the Milky Way. In a large enough cloud of sufficient density, with the atoms in random motion, condensed pockets of gas will arise which contain so many atoms that the condensation will be preserved by self gravitation. An immense number of atoms are necessary to give rise to such a contracting globule— about 10^{57}. Once such a condensation begins events happen very quickly—at least on the cosmical time scale. In a remarkably short time the globule, which originally extended for trillions of miles will have decreased to a few hundred million miles. The contraction continues until after about 27 million years and the internal temperature will have risen to about 20 million degrees and the pressure to several thousand million atmospheres. At this stage the thermonucler transformation of hydrogen to helium takes place and this release of nuclear energy generates sufficient outward pressure to halt further collapse. We then have a star, like the Sun, in a long term condition of stability for a few thousand million years, in which for every second of time some 564 million tons of hydrogen are being transformed into 560 million tons of helium.

Although this outline of the history of the birth of a star is based on the results of computerised calculations there is compelling observational evidence to indicate its essential validity. First, the protostars in the gas clouds of the Milky Way are evident in the photographs taken with modern telescopes. Then we have the remarkable photographs taken from the Lick Observatory of a section of the Orion nebula where the stars are forming, in which even over a period of a few years additional protostars become visible, and finally the recent infra-red studies of the sky show many objects, not seen in the visible region of the spectrum.

Our picture then, of the emergence of a planetary system, is entirely different from that presented to the British Association by Sir James Jeans less than half a century ago. His tidal encounter theory was presented as an event involving the close approach of two stars—sufficiently rare to be regarded as endowing the solar system, and the emergence of life on Earth, with a high degree of uniqueness. The contemporary view is that stars are born from the gas clouds of the Milky Way in large numbers and in this process the remnants of the nebulae will be retained around them. Thus we expect planetary systems around stars to be a common feature in the universe, and it will not have escaped your notice that it is in these identical nebulae that astronomers have recently discovered the complex molecules essential for organic evolution.

REACHING OUR ORBITING NEIGHBORS

The convergence of these two investigations has understandably been a great stimulus to those who seek for evidence of extraterrestrial existence. We have those who use radio telescopes to search for communications from such extraterrestrial beings as might have developed on planets around the stars. My personal judgement on the worthwhileness of such effort with our present equipment and understanding of the problem is reserved. My concern at this moment is with the more opportune and practical issue of the exploration of our own planetary system—the search for a deeper understanding of the processes of evolution and for evidence of extraterrestrial life forms within our own system. The possibility for these investigations by close inspection or contact with the planets emerged as soon as the Soviets succeeded in launching the first Sputnik on October 4, 1957. Almost immediately the efforts of the Soviet Union and the United States of America were directed to landing instruments and men on the Moon and to the despatch of space probes to the planets, especially Venus and Mars. Familiarity with these enterprises should not blind us to the magnificence of the achievements. The Russian success in hitting the Moon with Lunik 2 in September 1959 was a stark reminder of the dramatic advance of science and technology in the Soviet Union. The American response, culminating in the landing of Armstrong and Aldrin on the lunar surface 10 years later, will surely emerge in the perspective of history as one of the great triumphs of human endeavour. The return of lunar rocks to Earth has presented scientists with a means for studying the problem of the origin of the Moon and the early history of the solar system, far beyond the wildest dreams of our predecessors.

While this drama was unfolding the Soviets and Americans were striving to send space probes beyond the Moon to the planets. The immense dividends for science and national prestige provided an incentive to overcome a seemingly impenetrable barrier of engineering and technical problems. It is hardly surprising that when these attempts began in 1960 there were many failures. The atmosphere of the age was epitomised by the launching of the Soviet probe to the planet Venus in February 1961 at a time when the scale of the solar system was so inaccurately known that, however perfect the guidance of a rocket, insufficient data existed to be certain of a successful encounter. Soviet scientists had to rely on a radar measurement of the distance of the planet using the same radio telescope which they had hurriedly built to track the probe. This probe and many other Soviet and American probes failed to achieve success in these early attempts to reach the planets. Eventually the major problem of reliability, attached to these prolonged flights in space of many months duration was overcome. Just 10 years ago the Americans received on Earth photographs transmitted from a spacecraft which had passed within a few thousand miles of the planet Mars. The surface features of the planet shown in these photographs and those obtained subsequently seemed to reveal a hostile environment not conducive to any form of organic evolution as many had anticipated. By 1971 it became possible to place a spacecraft in orbit around the planet and during 1972 over 7000 pictures were transmitted to Earth by Mariner 9. Amongst them were high definition photographs with a resolution of 100 yards, that is 500 times superior to the best photographs of the planet hitherto available through telescopes on Earth. These photographs show mountain peaks as high as Everest, great canyons six times as wide and twice the depth of the Grand Canyon in Arizona, and sinuous valleys with branching tributaries resembling the watercut gullies on Earth. The American analysts who have studied these photographs conclude that these are, indeed, dried river valleys, that Mars is in an ice age and that the water is locked in the polar caps.

The perpetual clouds which cover our other near neighbour in the solar system—the planet Venus—encouraged our predecessors to the view that the planet might be rather similar to Earth—an abode for vegetation or perhaps organic life. Twenty years ago there were many learned judgements on the nature of this planet, but we did not even know how fast it was rotating on its axis, and could not be certain whether the surface was favourable for evolution, or was a boiling ocean or an arid desert. The Soviet probe Venera 4 settled these problems in 1967. As the rocket reached the vicinity of the planet, a capsule of scientific instruments was released, and this, slowed by a parachute

25

system, transmitted data to Earth as it descended through the atmosphere of the planet. Since that time a series of similar Soviet probes to Venus have confirmed the astonishing findings of that first investigation. The clouds are not of water vapour like terrestrial clouds. The atmosphere of the planet is 97% carbon dioxide. The surface conditions are extremely hostile, the pressure is 100 atmospheres and the temperature is above the melting point of lead and the boiling point of mercury. It is strange that Venus, in many respects so physically similar to Earth, should differ in this way.

Tantalising questions face us today. Has there been any kind of organic evolution elsewhere in the solar system, at any time in its evolutionary history? Is Mars really in an ice age, and if so has any kind of primitive organism evolved during its favourable epochs when water has flowed in the sinuous valleys? Why did Venus and Earth, differing only 3% in size, 18% in mass, and moving in reasonably similar orbits around the Sun pursue such radically different evolutionary paths? The answers to these problems are vital to the general problem of exobiology. Astronomically there seems to be compelling evidence that the solar system is not an unique planetary system. But given very large numbers of planetary systems in the universe what are the chances that evolution can occur? Do the recent discoveries about Mars and Venus indicate that the environment is so extremely sensitive that conditions suitable for evolution are improbable even in a billion planetary systems? I pose these questions at an exciting moment. As I speak, two Soviet probes are on their way to Venus and two American spacecraft to Mars. The Soviet spacecraft will reach Venus in October. They are 4 times heavier than the previous Venera probes, and it is widely conjectured in the West that a landing of equipment will be attempted, which may be designed to survive for some time the hostile environment

26

of the planet. The American Viking spacecraft to Mars, launched only days ago, will be in orbit around the planet and ready for landing on the bicentenary anniversary next year. The one ton lander is one of the most sophisticated miniaturised scientific laboratories ever devised. The primary objective is to search for life on the planet. A 10 ft arm will scoop soil samples into the body of the lander where they will be subjected to a variety of analyses to determine their organic chemical and microbiological content. The results from these particular probes may be indecisive—one has only to imagine the situation in reverse and envisage the difficulty of deciding where to land spacecraft on Earth from a distance of a few tens of millions of miles on the basis of photographs with a 100 yard definition. Even so, these flights to Mars and Venus represent a stage in man's technical accomplishment which will surely soon lead to a more decisive assessment of our place in the universe.

THE HUMAN MENACE TO MANKIND

It is therefore a sad and ironical reflection that we have been brought to this stage by the use of devices which are themselves designed to destroy mankind. The Soviet Sputnik was *not* launched into space by a rocket designed for this purpose. On 26 August 1957 Mr Khrushchev announced to the world that the Soviet Union had tested successfully a super-long-distance intercontinental ballistic missile and that the results showed that it was "possible to direct missiles into any part of the world". Six weeks later, this, the world's first intercontinental ballistic missile became the launching rocket of the first artificial satellite of the Earth. Of course, the United States was also developing the intercontinental missile, and three weeks after the launching of the Sputnik, Von Braun was at last given authority to

27

modify the Army's Jupiter ballistic rocket and it was this which launched the first American Explorer satellite on 31 January 1958.

It was no peaceful endeavour, either in the USSR or the USA, which gave man the power to launch scientific instruments into space and it serves no purpose to imagine that the space activities of these two countries today are innocent of military interests. The duality of the launching rockets needs no further emphasis. Even in the payloads the military interest remains dominant. The Senate papers which are readily available in Washington reveal that two thirds of the American payloads placed in orbit are under the control of the Department of Defence. According to the public releases of the United States intelligence services, since Sputnik One, 18 years ago, the Soviet Union has launched 834 space missions of which 516 have been for military activities.

East or West the picture is common—an almost non-existent dividing line between those activities which may either lead to the greatest human disaster since the calamitous Shen-Shu earthquake in 1556, which wiped out a million human beings in a few seconds, or which may, on the other hand, mark a profound intellectual advance in the development of civilization. We stand, perhaps uniquely, at the centre of this immensity—What is Man "who sees and fashions for himself a Universe".

I turn now to my second consideration—whether there is a limitation to human purpose in the adventure of understanding. Since this adventure has inevitably been a driving force in the advance of civilization I am concerned that certain fundamental difficulties exist, and that we must seek a new insight into the nature of human purpose. I appreciate that personal ambition sets a daily target to our localized purpose in life. The difficulty for society arises when this dominates existence to the exclusion of the search for the meaning of civilization. Such dominating localized ambition, be it national or personal, is destructive. Naturally it leads to human conflict, personally it leads to those forces which disrupt society of which we have abundant evidence today. It is, therefore, of great significance that we should attempt to discover why the deeper ambitions for the understanding of human purpose no longer exert such a dominating influence on our lives as they did for our forefathers. The Puritans sought for understanding by running from one hour long sermon to another. But for our generation science, coupled with technology, became the God through which we would find the road to economic and intellectual salvation.

THE ORIGIN AND EXPANSE OF THE KNOWN UNIVERSE

It is a vast region for debate and tonight I direct my attention to a single aspect. It is this. Throughout the whole of recorded history a consistent thread has been the intellectual purpose of man to discover the nature of the Universe. Today we refer to this as the cosmological problem, that is how did the Universe come into existence and what is its future? Hitherto, man has attempted to give either a theological answer or to believe that the solution would be discovered by scientific observation alone. Is the answer transcendental or material? Is the pursuit of science the only appropriate human activity for this search?

Over the centuries this problem has led to great intellectual drama in which philosophical conclusions have often been in advance of scientific observation. In the middle of the 18th century Immanuel Kant, in his "New hypothesis of the universe", enunciated the theory of the island universes. So did the English instrument maker, Thomas Wright, a few years earlier. The philosophical conception of extragalactic nebulae existing throughout space, was clearly stated at that time. But not until 50 years ago, when a means had become available for measuring these great distances, did the scientific community accept the fact that the Milky Way was not the totality of the universe. The contemporary argument as to whether the universe ever did have an origin or whether it is, and always has been, in a state of continuous creation has been clarified. But this has faced us with an imponderable conceptual difficulty.

May I first remind you of the observational features of the universe as they are revealed today. Our local environment is determined by the Milky Way system of stars. It is a galaxy of about 100,000 million stars in which our Sun is an average main sequence star about 5,000 million years old and with a future life expectancy of another few thousand million years. This galaxy is one of a local group consisting of the spiral galaxy in Andromeda, 2 million light years distant, and about a dozen other major galaxies. Our own local group is a minor cluster of the universe. Wherever we look in space we find large numbers of these clusters. When we observe them we encounter the phenomenon of the red shift in the wavelength of the spectral lines, interpreted as a doppler shift and indicating that the universe is expanding. If you wish to reflect on our significance in the cosmos it may be salutary to look towards the constellation of Coma, hold a penny at arm's length and remember that you obscure from your vision a cluster of a thousand galaxies, 350 million light years away, and receding from us with a velocity of nearly 5000 miles per second.

In viewing these clusters like Coma we are still only at the fringe of our possible penetration into time and space. I emphasise *time*, because it is a most important consideration that as we look out into space so we look back in time. The light from the Coma cluster has taken 350 million years to reach us and our view of that cluster is, therefore,

30

as it was 350 million years ago. In other words as we penetrate into space, so we penetrate into the past history of the universe. Now the measurements of the red shift have revealed that the velocities of recession of these distant galaxies increase with their distance from us. This gives rise to the simple conception that, in the past, the material of the universe must have been in a localised, concentrated condition. Further, since we know the relation between the distance and the speed of recession it is possible to calculate how long ago this concentrated condition existed. The answer is about 10,000 million years.

MYSTERIOUS QUASARS

This concept, that the universe evolved from a dense concentration of primeval material some 10,000 million years ago has been a spur to seek for observational proof. Since, when we look out into space we look back in time, the observation of more and more distant galaxies should enable us ultimately to observe the conditions in the early history of the universe. The development of the new technique of radio astronomy in the years following World War II released a flood of optimism that success in this search was at hand. Over the last 20 years, first through the discovery of radio galaxies, and then of quasars, we have been able to look back into the past history of the universe for many thousands of millions of years. Today it is a relatively simple matter to direct a large radio telescope towards the heavens and record signals from quasars which may be more than 7,000 million light years distant and which are receding from us with velocities of more than 150,000 miles a second. If the universe evolved from a dense condition, the observation of these distant objects takes us back at least three quarters of the time to this condition and to the beginning of the expansion. Now the

31

ability to make this penetration into time and space seemed to provide the means of securing the observational proof we sought about the early condition of the universe. This was not so; the results were suggestive rather than decisive, and for two decades astronomers have been immersed in an inconclusive debate about the meaning of these observations.

Unexpectedly the conclusive evidence has come from a different enquiry—a classic example of the accidental observations which permeate the history of science. Ten years ago the scientists of the Bell Telephone Laboratories in New Jersey, tested radio receiving equipment of high sensitivity working on a wavelength of 7 cm. This equipment had been built for space communication tests across the American continent via Earth satellites and unexpectedly produced an astronomical result of supreme importance. From all parts of the sky the astronomers discovered that this equipment was receiving radio noise 100 times stronger than the expected noise level in the equipment itself. Their immediate claim that this microwave radiation was the relic black body emission from the primeval fireball of the universe seemed somewhat extravagant. However, in 1973 and again last year, critical measurements of this radiation were made at wavelengths of 1 mm and less, using equipment carried above the atmosphere by a rocket and in a high flying balloon.

Tonight, I am not aware of any serious criticism of the view that these astonishing results are of a highly isotropic radiation equivalent to a temperature of 2.7 degrees absolute—and that this is the relic we observe today of the radiation from the high temperature phase of the initial collapsed state of the universe, 10,000 million years ago. The implication is that the measurement relates to an epoch a second or so after the initial expansion from the condensate had begun, when the temperature was about 10,000 million degrees.

This is the clarification of the cosmological problem to which I referred, namely the striking observational evidence that 10,000 million years ago the universe was beginning to evolve from a dense concentrate of primeval material. I said also that this clarification had presented us with an imponderable conceptual difficulty, and it is to this problem that I now turn. May I first ask you to reflect on the content of this primeval concentrate containing the entire mass of material which now forms the universe of our observation. It is hard to imagine this in terrestrial terms. The Sun is a million times more massive than the Earth. The Milky Way is nearly a million million times more massive than the Sun. How many galaxies are there in the universe? Certainly far more than the 100 million within the field of view of modern telescopes.

The total content of the universe must be far greater than these 100 million galaxies containing more than 10^{46} tons of matter. Our present enquiry concerns the initial condition of this material in its primeval state 10,000 million years ago.

ZERO RADIUS AT THE BEGINNING OF TIME

Apparently we observe today a radiation which is a relic of the high temperature phase of the universe, perhaps within a second or so of the beginning of the expansion. What does it enable us to infer about the physical state of the condensate in the beginning? Was there space in which the embryonic universe existed? Newton would have answered without doubt. For him space was absolute in which bodies could exist. Indeed it seems a commonsense view, for when we look at the heavens do we not see stars and galaxies existing in a space which also exists? Although the laws of motion and gravitation inherent in Newton's concept govern our daily life, they are incomplete and

33

unsatisfactory when we try to use them to explain the dynamical condition of the universe. In 1916 Einstein overcame these difficulties by abandoning this concept of an absolute space in which bodies could exist independently. In general relativity the properties of space are determined by the bodies contained in the universe. The solution of the equations of general relativity provide us with expanding models for the universe evolving from zero radius at the beginning of time. In that general respect the theory is satisfactory, furthermore in other matters,—such as the bending of starlight grazing the sun's disk—the theory explains effects which Newtonian theory cannot account for. The great difficulty is that these evolutionary models for the universe inevitably predict a singular condition of infinite density of infinitesimal dimensions before the beginning of the expansion. In this, the theory confounds itself and erodes our confidence in the applicability of the laws of physics to describe the initial condition of the universe. Until recently it has been maintained that this might be merely a mathematical difficulty arising from the assumption that the universe was exactly isotropic. But now we know from the measurements of the microwave background radiation that the universe *is* isotropic. Furthermore some recent theorems have demonstrated that in general relativity, the effects of self-gravitation inevitably lead to isotropy and hence to the singular condition of the universe in its initial state.

To what extent can we imagine the singular state as a reality in physical terms? Today, physicists feel satisfied that the known laws of physics apply to the behaviour of fundamental particles to dimensions of about 10^{-15} cm, that is a thousand million millionths of a centimetre. If we can imagine a dimension vastly smaller than this, a decrease by another million, million millionths to 10^{-33} cm we reach an interesting dimension in our attempt to describe the beginning of the universe. This figure of 10^{-33} cm is the unit of length through which quantum theory and the gravitational constant must be related. It is also the size of the universe 10^{-43} seconds after the beginning of time. It is here that we reach the conflux of Eternities—a barrier beyond which no known laws of physics apply. The search for a unified theory encompassing the laws of quantum mechanics and general relativity, a theory which would encompass the phenomenon of the gravitational and electrodynamic fields, is for us today like the alchemist's dream of ancient times.

If ever, in the future, the solution is discovered, then we might feel that scientific thought had approached one stage closer to the conception of the infinities of the singular state demanded by the theory. It is an embarrassing situation for science. The great achievements of observational astronomy and those of theoretical physics, have separately led to the same concept that the initial state of the universe was one of infinite density. The transference from the infinities of density and size at time zero to the finite quantities encompassed by the laws of the physical world may lie beyond scientific comprehension. Does man face this difficulty because he has externalised the object of his investigation? Is there reality in these externalised procedures? What is man's connection with the universe of atoms, stars and galaxies? Today we cannot evade this deepest problem of our existence by an escape into philosophical idealism or realism. On the contrary we are forced to recognise that, although in our daily lives, we can investigate problems as though the object of our investigation existed independently of us, this is not possible when we search for answers in the depths of the natural world. Or at least, I should perhaps say that at present it seems unlikely that externalisation of these fundamental problems is justified. The essence of the problem concerns the interpretation of the quantum theory. For at least half a century some of the most profound scientific and philosophical thought has been directed to this issue. The question is whether there can be a consistent interpretation of the quantum theory without external concepts; without the interaction of living or conscious beings with the object of investigation. There can be no formulation in quantum theory concerned with only a single unique entity—and that seems to be the issue which concerns us when we consider the initial state of the universe.

MAN'S TOTAL INVOLVEMENT WITH THE UNIVERSE

Indeed, I am inclined to accept contemporary scientific evidence as indicative of a far greater degree of man's total involvement with the universe. The life which we know depends on a sensitive molecular balance; the properties of the atoms of the familiar elements are determined by a delicate balance of electrical and nuclear forces. These and the large scale uniformity and isotropy of the universe were probably determined by events which occurred in the first second of time. For example, it is a remarkable fact that the existence, even of stars and galaxies, depends in a delicate manner on the force of attraction between two protons. In the earliest moments of the expansion of the universe, a millionth of a second after the beginning, calculations indicate that the temperature was of the order 10 million million degrees and the fundamental particles of nature—protons, neutrons, electrons, and hyperons existed with radiation as the controlling force. One second after the beginning, when the temperature had fallen to a few thousand million degrees there was a period when the ratio of protons to neutrons remained constant for a minute or so. This was the critical period when the natural constants determined the ultimate abundance of helium to hydrogen in the universe. It is an astonishing reflection that if the proton-proton interaction were only a few per cent stronger then all the hydrogen in the primeval condensate would have turned into helium in the early stages of expansion. No galaxies, no stars, no life would have emerged. It would be a universe forever unknowable by living creatures. The existence of a remarkable and intimate relationship between man, the fundamental constants of nature and the initial moments of space and time, seems to be an inescapable condition of our presence here tonight.

WHAT OF THE FUTURE?

What of the future? Will the universe expand for ever to a final state where no further energy is available, or will it recreate itself by collapsing again to a singular condition of infinite density? We do not yet know the answer. We do not know whether the expansion is solely the result of the initial motion of the particles at the beginning of time or whether the expansion is determined by a cosmological force of repulsion—a force opposing gravitational attraction which appears as a consequence of some solutions of the equations of general relativity. Hence we do not know whether the universe contains enough matter to overcome by gravitational attraction the forces now driving the galaxies apart. There are some who are willing to interpret the present data about the deceleration of the distant galaxies as an indication that the universe will collapse, and that it is forever cyclical, successively evolving and collapsing to the singular state of infinite density. In this case we may be privileged to exist in a unique cycle of the total history of the cosmos where the delicacy of the balance of the constants of nature narrowly determined the possibility that a part, at least, of this cycle should be knowable.

SCIENCE IN THE FRAMEWORK OF SOCIETY

I return to my starting point. Physically and intellectually we stand at the centre of immensities. Science itself is

37

neither a magic wand nor a poisoned arrow. Neither do I believe, as I have done in the past, that it is neutral in its impact. Its deepest pursuits are inextricably entwined with human purpose and existence. In a strictly localised sense a community can develop its scientific activity to support a framework of society. The manner in which it does so is of vital concern because science, through technology, is an immensely powerful force for good or evil. The researches into the structure of the atom half a century ago marked a tremendous advance in man's search for knowledge. Within a few years the massive technological application of this newly acquired knowledge led to the weapons which destroyed Hiroshima and Nagasaki. While these weapons were in preparation the development of rockets to carry bombs led to man's triumphal progress in the exploration of space. Today the delicacy of the balance for good or evil which these devices establish paralyses the imagination. As in war, so in peace. But today the distress of the human spirit is enhanced not merely by the inequalities amongst the peoples of the world, but especially by our failure to achieve the integration of science to meet this challenge.

At least in these physical and material affairs of life the relation of scientific activity to localised human purpose is clearly defined. The vital question is whether the framework of society in which science is pursued can develop the ethical basis and moral purpose necessary to ensure that in our future progress we overcome the forces leading to decay and destruction. It appears that within the last century the transcendental view of the world "sky-woven and worthy of a God" derived through centuries of religious thought and activity has been abruptly eroded. Tonight we may recall that the British Association meeting of 1860 in Oxford was the scene of the famous confrontation between Huxley and Bishop Wilberforce, between science and orthodoxy. The battle lines have separated, but the mind of man

38

is adrift and the peoples of the civilised world derive their satisfaction from activities which are so often alien to, and destructive both of the physical and intellectual environment. One world view has been eroded and the inadequacy of its substitute is being demonstrated. We have deluded ourselves that through science we find the only avenue to true understanding about nature and the universe. Furthermore, we have persuaded the society in which we work to support our activities in the belief that our discoveries will inevitably, in some way, be of practical benefit. The simple belief in automatic material progress by means of scientific discovery and application is a tragic myth of our age. Science is a powerful and vital human activity—but this confusion of thought and motive is bewildering to man, and it is a most alarming thought that the present antagonisms of society to scientific activity may deepen further.

As a scientist I believe that observable phenomena are subject to scientific understanding. The pursuit of this understanding is an essential occupation of modern society. But I cannot believe that this quest embraces the totality of human purpose. We can apply the spectroscope to gain an understanding of the sunset, we can send the space probe to Venus, but we may never apprehend the ethos of the evening star. Human existence is itself entwined with the primeval state of the universe and the pursuit of understanding is a transcendent value in man's life and purpose.

Wherefore

God by wisdom hath founded the earth; by understanding hath He established the heavens.
                                    —Proverbs III, 19

Wherefore

An Inquiry Beyond Whence

Zev Zahavy

When I behold the heavens, Thy handiwork;
the moon and the stars which Thou hast established; what is man that Thou art mindful of him,
and the son of man, that Thou thinkest of him?
Yet, Thou hast made him but a little lower than
God, and hast crowned him with glory and honor.
Thou hast made him to have dominion over the
works of Thy hands; Thou hast put all things
under his feet . . . O God, how glorious is Thy
name on all the earth!

—David, son of Jesse

1

A COURAGEOUS QUESTION

One wonders how many of the distinguished scholars who
listened to the significant presidential address delivered by
Sir Bernard Lovell on that singular summer day in August
1975 recognized immediately its extensive ramifications.

43

The subject of the paper at the time of its presentation was pointedly designated, "In the Centre of Immensities." Its title relates to the classic work *Sartor Resartus*, by Thomas Carlyle, which first appeared in *Fraser's Magazine* in 1833–34, at about the time when England's leading men of science were advancing the cause of their newly formed organization, the British Association for the Advancement of Science. Among others, Carlyle boldly addressed science and expressed his concern for man's basic existentialist identity and destiny. He was prompted by a desire to develop meaningful relationships between man and the mysterious universe of extensive expanse.

It is my estimation that the purposeful selection of a quotation from Carlyle before a body representing the distinguished scientific establishment conveys more than merely a setting for cosmological analysis. I believe that it contains a hint of majestic proportions insofar as amending the current materialistic outlook and philosophy of science. It points the way for the assumption of a new posture by men of science in their encounter with problems of existentialist overtones, and this is somewhat implied by placing before them for serious consideration, Carlyle's inquiry, wherein he says of man, "Stands he not thereby in the centre of Immensities, in the conflux of Eternities?"

The New York Times deserves a measure of credit for exhibiting an alertness to the full significance of Professor Lovell's presentation. It was published with some slight revisions as the lead cover-article in the Sunday magazine section on November 16, 1975, where it was endowed with the simple, but extremely provocative title, "Whence," to which was further appended a rather enigmatic, philosophical subtitle, "We Are What We Know About Where We Came From."

Quite appropriately, then, "Whence" may portray a milestone in the maturing of modern scientific thought. The essence of "Whence" signals to the contemporary scientist

44

the explicit fact that other questions besides the normal "how" of the laboratory should come to the awareness of the researching mind. Perhaps this may come as a surprise to the current scientific generation, who, for the most part, are steeped in a materialistic attitude toward life and toward the tasks in their demanding discipline. Most of them, since their early freshman years, were indoctrinated with a scientific spirit that stressed the virtue of committing all investigations to the arena of "how." Now, thanks to the courage of a leading scientist in the British Association, the staid rank and file are suddenly confronted with the challenge "whence."

At first glance it may seem that the question "whence" may ruffle the traditional tranquility of the heretofore impregnable laboratory fortress of science and disturb its tidy security. After all, the query is a cardinal interpolation and points to an ultimate origin of some sort. "Whence" suggests the need to explore a philosophical abode where abstract concepts prevail. It anticipates the need to consider the broad domain of idealism. Science, however, is committed mainly to research in the world of concrete substance, and indeed the physical cosmos is now envisaged as extending into the remote areas of the atom's invisible nucleus as well as unto the farout reaches of space. Does "whence" imply that the candid scientist in the coming space age must inevitably turn his sights in the direction of the divine domain?

Let us further consider the implications of Professor Lovell's reference to an essential motif in *Sartor Resartus*. It is possible to discern in Carlyle's work additional parallels that are quite relevant to our contemporary turbulent era of advanced technology. Some of the problems viewed by Carlyle in his generation may currently apply to our own day and age, as well. Behind a facade of amusing satire and frivolity, *Sartor Resartus* enunciates some very earnest and prudent thoughts.

45

For example, Carlyle was moved to produce his masterful work as an expression of disdain for the extensive materialistic outlook of his generation. Like the prophet of old, he ranted against the spiritual deficiencies of his age. The book's title is commonly translated "Tailor Retailored," and the philosophical influence of German spiritual idealism is strongly evident.

According to Carlyle, civilization is a tired robe enveloping the essential world soul. Since appearances are deceiving, physical identities can hardly claim reality. The most important aspect of life is the divine principle, but it is concealed by the extensive garment of nature. While the godless ones experience negative points in a life bereft of spiritual values, the Godly ones endeavor to retrieve valid meaning from life's depths through dedication and spiritual heroism.

Life is woven with tragic elements such as the finitude of worldly dimensions dominated by time and space. Nature's garments symbolically conceal from man the true essence and meaning of the universe. God's divine spirit is hidden behind the splendorous vesture of creation. Contrary to popular belief, which sets happiness as the ultimate goal in life, Carlyle suggests that communion with God is a greater achievement. Life's enigmas are not readily discerned, nor can life's rewards be easily attained.

In reference to the science of his day, Carlyle notes that it had hardly penetrated the shrouded spiritual mysteries concealed behind the outer vestment of nature. Carlyle abhors the condition of man enslaved to custom. The seeker of truth must wage relentless battle against the futile elements of custom. Science, however, is enamored with custom, and in dogmatic fashion, it helps to maintain human bondage.

Sartor Resartus reflects Carlyle's own spiritual struggles to set a meaningful course in life. In some respects, his work evokes an existentialist mood. Carlyle rejects the assumption of a negative attitude toward life, since it would eradicate God and foster a hopeless existence. Instead, he announces his faith in God, and he endorses the possibility of reaching divinity through hard labor and courage.

If we may assume that Professor Lovell's citation from Carlyle is indicative of a sympathy toward its broader, general implications, then we are presented with a statement of far-reaching proportions. Let us examine at further length the text of the selected quotation. The passages appear in chapter ten of *Sartor Resartus*, which is endowed with the title "Pure Reason." This calls to mind Kant, who was one of the foremost discoursers on pure reason.

The paragraph containing the quotation commences in the following manner: "To the eye of vulgar Logic . . . what is man? An omnivorous Biped that wears Breeches." The famed exponent of logic is Aristotle. Carlyle here indicates a none-too-great affection for that peripatetic philosopher. Carlyle returns to the Platonic theme of his literary symphony, and continues his exposition with evident warmth.

To the eye of Pure Reason what is he? A Soul, a Spirit, and divine Apparition. Round his mysterious ME, there lies, under all those wool rags, a Garment of Flesh (or of Senses), contextured in the Loom of Heaven; whereby he is revealed to his like, and dwells with them in Union and Division; and sees and fashions for himself a Universe, with azure Starry Spaces, and long Thousands of Years. Deep-hidden is he under that strange Garment; amid Sounds and Colours and Forms, as it were, swathed-in, and inextricably over-shrouded: yet it is sky-woven and worthy of a God. Stands he not thereby in the centre of Immensities, in the conflux of Eternities? He feels; power has been given him to know, to believe; nay does not the spirit of Love, free in its celestial primeval brightness, even here, though but for moments look through? Well said Saint Chrysostom, with his lips of gold, "the true SHEKINAH is Man": where else is the GOD'S-PRESENCE manifested not to our eyes only, but to our hearts, as in our fellow-man?

47

The strains of idealism continue to flow from the philosopher's pen with delicate charm:

> In such passages, unhappily too rare, the high Platonic Mysticism of our Author, which is perhaps the fundamental element of his nature, bursts forth, as it were, in full flood: and, through all the vapour and tarnish of what is often so perverse, so mean in his exterior and environment, we seem to look into a whole inward Sea of Light and Love;—though, alas, the grim coppery clouds soon roll together again, and hide it from view.

Carlyle invokes a thoughtful mood of theistic idealism, which suggests the revelation of the Divine Presence, the *Shekinah*, so to speak, through an awareness of ego noesis; and a brief glimpse of the Divine Personality through the appearance of natural law from whence emanates the vibrations of dynamic moral norms. We shall discuss such concepts more fully in a later chapter. At the moment, let us briefly examine some aspects of idealism and materialism that prominently relate to the contemporary scene.

There are some who submit that basically philosophy is comprised of two principle systems: idealism and materialism. Idealism sponsors the view that mind or spirit is primary in the universe. Materialism proposes that matter is primary in the universe. More specifically, idealism looks beyond that which appears to common sense experience in search of an ultimate nonphysical abstract reality. It considers the concepts and values consequently emerging from such an exploration as the fundamental mainstay of the cosmos. Materialism, on the other hand, regards all such emergent notions as items readily reducible to material things and processes.

Idealism embraces a number of subordinate doctrines such as subjective idealism, objective idealism, and to some extent, pantheism. The latter would come under this heading by virtue of its opinion that only God, including His attributes, alone exists. For the pantheist, the material

48

world is either an aspect of God, or the entire appearance of God. As an aspect of God, some elements of idealism may be professed; however, if the universe is considered to embrace the entire appearance of God, then such a pantheistic notion could better serve the interests of the materialist. In all, idealism entails a divergent spectrum of classifications ranging from Platonism and panpsychism to personalism and absolutism. Our interest in the term refers to a thesis common to all elements of idealism. Perhaps we may offer as its identification the term "theistic idealism." In contrast, when we speak of materialism, we refer to what could be called "atheistic materialism."

The idealists reflect an essential aspect of Platonism by regarding ideas and ideals as prior to and fundamental for material construction. At the other pole, materialists consider ideas as a derivative of matter and of secondary significance, much as did Democritus, Empedocles, and Lucretius. An important point in absolute idealism is the emphasis upon relating the identity of reality with the Absolute.

It is tempting to depict the fluctuations in human intellectual history on a simplified scale, with a pendulum swinging between two integral doctrines. If one yields to such a description of intellectual variations, the two basic extremes deserving to be so cited are idealism and materialism. Generally speaking, then, the main contours of Western social and intellectual expression could be depicted as inclining toward either idealism or materialism. The type of idealism or the cast of materialism a generation chooses to reflect can be expected to influence the popular mores and social behavior of its society. For this reason, the promotion of theistic idealism could have far-reaching beneficial effects, since it subscribes not only to God, but also to a high moral code considered to be of divine essence.

Theistic idealism exalts God as the creator of a universe

beyond or outside of His own being. Although the material world is dependent on God, it is not an aspect or appearance of God. Beneath the banner of theistic idealism a metaphysic becomes possible that may favorably synthesize religious doctrine and belief in accord with its principles. Theistic idealism upholds God as the fundamental, perfect creator of the universe. Theistic idealism does not necessarily dismiss the physical cosmos as an illusion of the mind. It does regard the material world as conforming to laws and formulae that preceded all existence.

Idealism traces its roots back to Plato, whose "Doctrine of Ideas" exalted the Idea or Form as being more real than its actual material counterpart. The Idea described a universal as permanent in contrast to its particular, temporal counterpart. When the particular conforms to the universal, it can only approximate the perfection of its formula, blueprint, or design. That which is real for Plato must be eternal, indestructible, and intangible. The abstract idea, as apprehended by the intellect, fulfills these requirements. Medieval philosophers established the Ideas as paradigms for divine creation, and they therefore were considered to exist in the divine intellect.

The being of all data that is experienced by the senses is only transitory in its nature. Perceived matter is temporal, variable, subject to the vagaries of time and tide, and therefore impossible to identify at any given moment as the permanent embodiment of its species or class. The Real, on the other hand, exhibits an indubitable permanence, because it serves as the law to which matter is committed. Such law is enduring and immutable. It is beyond the tangible reach of the senses. The Idea as the Real is uniquely self-subsistent; it is dependent neither upon the mind nor upon the material world for its existence.

First there was the Idea or Form, then there followed the implementation of the Idea through the appearance of

50

matter. The objective of matter was to subscribe to the law, the formula, or the equation to which it was committed. Matter itself could not compose its own paradigm, nor could it propose its own Idea. The pre-existant Idea determined the manner of particle composition and behavior. Even the erratic quantum qualities act in accordance with their pre-existant Idea, which endorses their erratic motion.

But, from whence came the Ideas themselves? Their source derived from what Plato identified as the "Good." Perhaps an acceptable explanation of this concept would simply be God.

When Carlyle espoused the spirit of Platonism in line with German idealism, he was inspired by the recognition of a universal as the supreme entity in the cosmos. The idealist notes that matter itself is committed to an ideal, namely its Idea or Formula. For the materialist, matter is responsible to nought but itself. In a civilization inspired by idealism, man recognizes the supremacy of a higher authority. In a self-serving, materialistic society, man assumes that he is only responsible to himself.

The grand tradition of idealism occupies a distinguished chapter in the history of British thought. George Berkeley (1685–1753) developed what he termed "immaterialism" in an age of English empiricism, when the doctrines of the determined materialist, Thomas Hobbes (1588–1679), the empirical dualist, John Locke (1632–1704), and the empirical skeptic, David Hume (1711–76), were competing for the favor of British acceptance. Berkeley considered that man can only know his own ideas. He upheld the concept that all we know are sensations and ideas, and considered a proof for the existence of outer material substance as unlikely. The world of ideas was paramount, and he conceived two varieties, namely: ideas within the mind wholly; ideas that come to us from without, we know not whence-sensations. Since there are no material substances, the cause must be incorporeal. We assume that our ideas belong to our spirits, so these outer ideas are similarly in the custody of a "spirit," who is better identified as God.

In Germany, idealism blossomed along several hues. Gottfried Leibniz (1646–1716), a rational idealist, sponsored a metaphysical idealism, contending that reality consists of monads that affect each other. He proposed a series of realms of being. God is the supreme, uncreated spiritual source. Created substances are immaterial, and the self-conscious members are formed in God's image.

The towering figure of Immanuel Kant (1724–1804) arises in the age of Enlightenment. It was a period that glorified knowledge, extolled the sciences and the arts, and encouraged civilization and progress. The Enlightenment in England was somewhat slower in its development and not as radical. Nevertheless, the English influence is strongly apparent, with the ideas of Locke practically formulating the whole spirit of the Enlightenment. Generally speaking, the Enlightenment developed the scientific view of the material world, and absolutized scientific knowledge.

Often overlooked is the English influence on the development of German idealism. Toward the middle of the eighteenth century English thought was transmitted for German study through the translations of Locke, Hume, and the English moralists, Shaftesbury, Hutcheson, and Ferguson. Consequently, German philosophy assumed an eclectic disposition, with emphasis on the rational, teleological aspects of the universe and mankind's history. Reason was applied to remove the blemish of popular superstition. A rational theology developed, emphasizing nature, so that when Kant brought forth his transcendental idealism, he reflected the spirit of the times.

Kant was intrigued with English empiricism, and it motivated his own philosophical thinking toward the contemporary issues of his age. His purpose was to diminish the skepticism of Hume and to eradicate the specter of materialism, fatalism, and atheism. Kant's philosophy is enormously complex, so that we shall suffice by mentioning only a few of its highlights. Hume's skepticism moved Kant to distrust physical science as a total explanation of knowledge. Kant adopted a Neoplatonic position insofar as conceiving a suprarational self through which an ethical motif may gain ascendency. He regarded knowledge as universal and necessary. In his analysis of pure reason, he concluded that the will and not reason is decisive in determining things. Practical reason is superior to theoretical reason. Religion within the bounds of reason is exemplified in a high morality. The moral law is a categorical imperative.

The spirit of absolute idealism is reflected in the writings of Johann Fichte (1762–1814), Friedrich von Schelling (1775–1854), and Georg Hegel (1770–1831). The problem for these post-Kantian idealists was the search for a common denominator for the purpose of unifying the systems of knowledge embracing nature, science, morals, and aesthetics. It was desirable to solidify the various tendencies into a systematic form.

Kant had left a lasting impression on his successors. In his opposition to the naturalistic world view with its mechanism, atheism, and hedonism, Kant limited natural science to the field of phenomena. There was, he concluded, a higher type of truth than that offered by scientific facts. What Kant called *das Ding an sich*, "the thing-in-itself" or noumenon, remains beyond the reach of sensual identification. As an abstraction, it becomes a necessary idea of reason, and a regulative principle desiring a unification of the soul, the world, and God. Within man, the cognizance of moral rectitude implies the existence of a supersensible world, and this is closed to the physical methods of research. This moral law is Kant's categorical imperative. The mind possesses concepts and presuppositions that are useful in assessing the world. It is not a question of reflecting upon cosmic phenomena, but endeavoring to understand and interpret them. Man's concepts provide him with the tools for interpreting, by applying the principle of synthesizing.

Kant's philosophy found favor in the eyes of the new generation. By minimizing its claims to knowledge, it offered an opportunity for turning from the natural sciences as the predominant influence in life. Along these lines, Fichte, Schelling, and Hegel commenced their inquiries with the intelligible world, or freedom, emanating from the moral law. The ideal or supersensible world, the world of the mind or spirit, was installed as the real world. All knowledge and experience was considered to flow from self-determining spiritual expression, and with it the attempt to solve humanity's problems became more conceivable.

Mention should be made of Friedrich Schleiermacher (1768–1834), who was a distinguished theologian and an essential spokesman of the German idealistic movement. He sought a concept of reality that would be acceptable to both the intellect and the feelings. Schleiermacher turned away from Fichte's view that considered the Ego as the source of all reality. Instead, he assumed the existence of a real world, and he inferred a transcendent basis for all thought and being. Since our perceptions are not equipped to gain sufficient knowledge concerning the original source of things, it is necessary to seek the absolute principle, and know the identity of thought and being. God is this principle. He is the absolute unity or identity of thought and being. Schleiermacher endeavors to harmonize elements of pantheism with dualism by identifying God and the world as a unity. Although God and the universe are inseparable, things and the world have a relative independence. God is a spaceless, timeless unity. The world is a spatial-temporal plurality.

The preceding constitutes the basic concepts of German idealism that Thomas Carlyle helped to introduce upon English soil. He was ably joined in this project by Samuel Taylor Coleridge, William Wordsworth, and John Ruskin. It is strange that these four personalities are popularly known for their literary contributions, but hardly at all recognized for their philosophical pursuits. Chiefly representative of a subsequent school of English idealism are Thomas Hill Green, Bernard Bosanquet, and F. H. Bradley, along with Edward Caird, John Caird, and James Ward.

Thomas Hill Green (1836–82) sponsored an objective idealism that sought to supplement natural science with a spiritual metaphysic. Utilizing Kant's criticism, he turned upon the popularly viewed concepts of empiricism and utilitarianism. Specifically, he attacked the empiricism of Hume, the hedonism of Mill, and the evolution of Spencer; and he adduced as a common major failure their assumption that phenomenon is a product of itself. Although man is a biological phenomenon, he also possesses spiritual qualities, and it is the spiritual principle in man that makes knowledge possible and morality meaningful. It is not possible to derive a purposeful knowledge of nature without a unifying spiritual principle. The intelligence of man makes it desirable for him to transcend nature. Man may apply his will toward realizing the idea of the self.

Bosanquet's idealism stressed the point that every aspect of finite existence must transcend itself. Through such a process, it becomes possible to turn to other existences. Subsequently, the existing particular may confront the whole. This concept reflected the general principle, which gained wide popularity in the early twentieth century, that philosophical truth was an all-embracing unity.

Francis Herbert Bradley (1846–1924) took up the cudgels of Green in his battle against empirical and utilitarian assumptions. In line with the German idealists, Bradley upheld the importance of metaphysics in the search for truth. His conclusions followed the patterns of Hegel and Kant. Man is impelled to reflect upon ultimate truth, and his knowledge of the Absolute is certain; however, it is also incomplete. The ultimate reality is a self-consistent whole embracing all differences in an inclusive harmony.

Once more returning to Bernard Bosanquet (1848–1923), who influenced Bradley and was influenced by him, we note that he disagreed with the latter on several points. Bosanquet stressed the adequacy of thought as a means for reconciling immediacy and logic. He also identified a concrete individual or whole in higher synthetic experiences, and he further conceived the existence of a collective will.

Green, Bradley, and Bosanquet, as the foremost spokesmen of objective idealism, sounded a call for universal harmony, wherein the organization of experience could be welded into a living totality and systematic whole, thereby healing all conflicts, unifying all differences, and harmonizing all discords.

Having cursorily identified the allied forces of idealism, let us pause momentarily to glimpse some of the notions expounded by the proponents of materialism. The materialists assert that the essence of the real world is limited to the material elements therein, as they appear in various states and relationships unto each other. In their view, only matter exists. The mind or spirit is dependent upon reality for its operation and function. Since matter is the subject of science, the many states of matter become the object of scientific inquiry and evaluation. The mind and all ideas are subject to matter. Conscious perception and all of the uniquely human functions, such as emotion, ambition, and desire, are excluded from the serious concern of the materialist, since they do not appear to be properties of matter. Matter has no psychological backbone, and neither souls, nor spirits, nor gods exist, since they are conceived as divorced of matter. Everything that appears or occurs in the universe is the consequence of some antecedent physical condition.

In this respect, the proponents of materialism turn out to be the staunchest supporters of a determinist doctrine, yet recently, some materialists have turned away from determinism, particularly in consideration of the enigmatic quantum behavior. Science, however, is largely favorably disposed toward a materialistic doctrine, because most of its analysis involves matter, and its basic methodology relates to applicable physical situations. Materialists consider that their views serve science best, and to reinforce their assumptions, they point to the progress science has made in explaining the physical nature of the world through a program of investigation based upon materialistic principles.

The ancient forerunners of materialism were the Greek thinkers Democritus, Empedocles, and Epicurus. Generally speaking, the materialistic position described above well reflected their views. The famous Roman Lucretius was motivated along similar lines of materialism when he wrote his well-known piece *De Rerum Natura*. During the period dominated by Aristotelianism and the Church, the voice of materialism remained at low ebb. With the coming of the Renaissance its theme was renewed. Thomas Hobbes appeared as its most vociferous patron. Quite simply stated, Hobbes advocated the notions that the mind is a brain substance; images and ideas are motions in the brain; and the whole universe is particles of matter in motion. He further postulated that incorporeal substances cannot exist, and he rejected angels, the soul, and religion's God.

We may also take note of an interesting materialist and a contemporary of Hobbes who sought to harmonize Epicureanism and Christianity. Pierre Gassendi (1592–1655) subscribed to a materialist interpretation of the universe by upholding the supremacy of matter in the whole physical realm. Yet, he conceived God as the creator and director of the cosmos, and he permitted man an immortal intellect apart from his corporeal soul.

The advancements made in science, especially in chemistry and biochemistry during the early nineteenth century, resulted in an increased support for materialism A compelling impetus on its behalf was the appearance of Darwin's *Origin of Species* in 1859, and his *Descent of Man* in 1871. Materialism welcomed the sanction of one of Darwin's tenets that survival of the fittest was an impersonal trait of nature, beyond the reach of any immanent power, and bereft of any transcendent purpose. It was further strengthened by Darwin's assessment that man was nothing more than a biological entity at the end of a meaningless physical chain. From that point onward, the swift advance of materialism could not be curbed. Contemporary materialism holds sway in all walks of life, on all levels of existence, and dominates the academic and cultural environment of the human family. Whether it be in the sphere of science or philosophy, psychology or technology, materialism reigns supreme.

It is no little wonder, then, that a pro-idealist quotation emanating from a prominent scientist in the late twentieth century should command widespread interest and attention. Sir Bernard Lovell assumes a courageous position of leadership in what may yet develop into a resurgence of space-age idealism on all fronts of human endeavor. It is not easy to be a pioneer or forerunner in modern society. The strong British materialistic tradition of the late eighteenth and early nineteenth centuries is sustained by the influential power of Locke and Hume; the psychological-epistemological theories of the British School; Bentham's Utilitarianism; and Comte's Positivism. Add to these forces Mills's skeptical empiricism; Darwin's theories; Spencer's cosmic evolution; Haeckel's monistic philosophy of nature; and Ernst Mach's new positivism, and one is exposed to a formidable array of anti-idealistic sapience. For over a period of almost two centuries science has been mainly committed to a mainstream of atheistic materialism, and the ethics that such a viewpoint could tolerate for society may dismally reflect, as Spencer taught, a hedonistic and base utilitarianism.

It is therefore important to gather together like-minded parties to support and further endorse the view that Professor Lovell has boldly brought to the attention of the British Association and the world at large. If we interpret Professor Lovell's Presidential address correctly, he is pointing to the timely need for a neoteric transfer from the heretofore rigid scientific commitment to atheistic materialism unto a more flexible position that could tolerate the basic principles of theistic idealism. If this be indeed the case, then "Whence" assumes the stellar quality of a historic declaration. It calls for the sighting of new directions by the scientific establishment; for a progressive, new era in its scope of purpose; and an upward adjustment of its attitude in regards to the ultimate destiny of man in his universe. An invitation to scientists to consider the significant rewards that may accrue by seeking new corridors of thought in theistic idealism is surely a momentous event at this critical point in human history.

Professor Lovell should take heart in pursuing such a course, because he reflects the grand philosophical tradition of British idealism. Indeed, he travels in the highly respectable company of such distinguished scholars as George Berkeley, Arthur Collier, Thomas Hill Green, F. H. Bradley, and Bernard Bosanquet. Of course, we must not overlook the fact that he treads a path that was once heroically outlined by the fearless pen of a Thomas Carlyle.

If the full message of Carlyle's classic work is not only implied, but moreso, conscientiously taken to heart, then possibly the time may be near at hand when bold scientists will express a desire to penetrate beyond the substantial veil of nature's physical formulae, to seek truths of nobler existentialist significance. Perhaps, like Carlyle, the scientist may now be prepared to escape from the custom of a bland servitude to a rigorous materialism, and soar unto the ethereal heights of idealism in search of more meaningful explanations for the existence of the universe, its generation of organic life, and its gifted intellectual product, man.

But, what is the consequence to scientific thought if such a search leads to God? Must then the scientist, who has been bound by custom to an iron-clad atheistic materialism, hang his head in embarrassment, and burrow his way back to the world of physical "garments" and laws, where he has permitted a widespread agnosticism to prevail, and from that nether point of concealment deny that God exists? Or is it possible that, out of the depths of despair to which civilization has fallen, the need for hope and promise becomes an existentialist imperative, so that it may even behoove the modern scientist, as Carlyle was so moved, to endorse the possibility of reaching divinity through hard labor and courage?

These and other enquiries which we postulate in the following folios come to mind hard on the heels of the courageous question "whence," which seems to suggest the need for a new type of idealism to which modern man may become committed. While Professor Lovell may harbor some sympathy for the high-minded principles of traditional English idealism, he does not seem to find any of its particular representations as potentially applicable to current situations. Perhaps the classical spirit of philosophical idealism and realism sufficed for a preatomic civilization; something more appealing may be needed for the perplexed generations of a coming space age, as Professor Lovell states, "Today we cannot evade this deepest problem of our existence by an escape into philosophical idealism or realism."[1] If the old idealism and realism are insufficiently endowed to service a modern society, what other intellectual direction appears as an alternative? Considering that certain elements of idealism are desirable and useful, can they be updated and wedded to a meaningful existentialist accommodation? Does an existentialist idealism offer a vision of promise for the future? Let us examine this in the ensuing chapters.

2

ASPECTS OF SCIENCE AND THEOLOGY

Without doubt, the prodigious achievements of science merit unequivocal acclaim and unreserved approbation. Science has constructed a magnificent edifice of knowledge in which nature and the universe emerge as a panorama of ordered cohesion. In its persistent search for the truth, science has endowed man with a multitude of blessings ranging from technological marvels of everyday life to a stunning array of beneficial products for human survival. The dynamic enthusiasm of scientific research has brought man to the threshold of nature's innermost secrets at the core of the atom, and on the distant horizon turns his keen intellect to the challenge of the boundless universe itself. There is no doubt but what science has wrought for man is the single greatest wonder in the world.

But what of the spirit of science that effuses from the disposition of an all-too-human scientist himself? Despite its grandeur as a mighty discipline, science is, after all, an effulgent dimension of man's mind as he conceives his environment. Cohesively considered, science is but one aspect of mortal expression. The success of science may be credited to the artful application of necessary methods and principles of research in accordance with long-established

---

1. Sir Bernard Lovell, Supra, p. 35.

immutable laws of nature. Simply stated, the scientific method is man's inductive response to nature's bidding. Of course, we do not imply through such simplification that the scientific method is subaltern as an expression of human ingenuity and brilliance. In fact, it remains uncontested as one of the few great creative achievements of the human mind. What is suggested is that man could not have produced the complex structure of a scientific discipline had nature exhibited a limited pattern of elementary principles of operation, rather than its abounding enigmatic structure of inscrutable intricacy. It is to nature's credit that man has composed the multiplex, composite system of science. If nature were harnessed to a crudely simple design, science would have been quite unsophisticated, since the universe would have been less fanciful.

Science is an expression of the human mind in much the same fashion as is philosophy or theology. It represents one of the attitudes that man may assume as he gazes out at the great expanse of nature. The scientist regards his environment from a technological frame of reference. He is concerned with the structural aspects of matter and existence. His analytical gaze sweeps from the vastness of the extragalactic regions in outer space to the subatomic particles in the microscopic confines of nature.

The discipline of science requires objective observation and systematization leading to classification. The classifications are further refined and through a process of inductive or deductive reasoning, often involving further experimentation, general laws of natural behavior may be established. An intriguing feature of scientific research is the ever-increasing availability of new information for the inquiring mind. The universe appears to contain an unending wellspring of substance and relational variations. Man never seems to be able to satiate his voracious appetite for further knowledge in his pursuit for scientific truth, because at each new turn of the hunt, unexpected principles emerge to further astonish him.

A convenient definition of "science" is to consider it as a process of research for the purpose of developing a specific body of knowledge relating to the structure and behavior of matter in the universe. The acquisition of such a doctrine is usually accomplished through the application of systems elaborated and refined by former generations. By applying such an explanation it becomes possible to identify *tradition* in scientific development. Tradition is an important element for the promotion of a scientific heritage. Once the scientist recognizes the value of tradition for the expansion of scientific knowledge, the appearance of such a sentiment elsewhere in intellectual disciplines should not be blithely disregarded as irrelevant and invalid. If the scientist understands and respects the traditions of science, then he should strive to understand and respect traditions in other developed orders of thought, such as philosophy and theology.

The information that the scientist seeks relates to that phase of matter which may be identified on the physical level of phenomena. He is not concerned with the problem of a possibly far-reaching, unseen reality beyond the immediate sensual scope of his investigations. In any case, the information that science contributes may be applied to a further examination of the reality of knowledge by disciples of philosophy or religion. The scientist himself, however, does not directly become involved in such a topic of research. This is not to say that the scientist must ever avoid contact with abstract concepts. In fact, science very handily utilizes abstract symbols and references in many aspects of its operation, particularly in formulae and mathematical equations. Such applications, however, mainly relate to a demonstration of concrete parallels in phenomenal categories. Consequently, symbols in science appear more as a convenient operational extension of the world of matter than as strictly abstract contemplation.

Although the term *science* conjures up an image of high respect and incontrovertible authority, there is a human factor involved that has a strong bearing on the direction and conclusions science ultimately adopts. In the process of investigation, the scientist, notwithstanding his commitment to objectivity, arrives at a point of judgment wherein the human factor controls vital decisions. Certainly, the individual who is endowed with a higher quality of intelligence will probably invoke more meaningful choices, and therefore appear as the better scientist. In many respects, the scientist must render judgments no less significant to his discipline than the judgments that the philosopher or the theologian must make in their respective disciplines. Moreover, the scientist must realize that his knowledge and experience in his own discipline does not automatically grant him the warrant to pass judgment on matters outside the periphery of his expertise. It is not that the scientist is any less intellectually proficient than the philosopher or the theologian. The fact simply remains that while the scientist recognizes the need to follow a carefully delineated course in arriving at scientific conclusions, he may sometimes be tempted to make judgments in alien areas merely on the basis of personal bias or opinion. The proper scientist would hardly think of dispensing a scientific judgment unless he had carefully abided by the accepted procedure of observations leading to an hypothesis, after which the validity of the hypothesis would be re-examined so that it may be used as a reliable basis for further observations. This is a lengthy procedure for arriving at a particular judgment. No less serious an attitude should be expected in the research field of philosophy or theology, although its approach may vary from scientific methodology. The scientist, therefore, should be quite consentient to the notion that when he lacks the formal background of the philosopher or theologian, he should hesitate to express a hasty or unfounded judgment in those areas.

A popular view considers modern science as developing with the appearance of Isaac Newton's *Philosophiae Naturalis Principia Mathematica* in 1687, although its full impact was delayed until Voltaire's essay clarified the Newtonian system in 1737. Newton's thesis influenced the scientific method for several centuries in that it provided a suitable procedure for objectively considering and working with the physical properties of matter. In an age of prolific discovery, he taught the usefulness of scientific demonstration. It is interesting to note that Newton, who is credited with being the father of modern science, was hardly averse to the principle concepts of seventeenth-century theology.

To some extent, Newton may also be identified as a progenitor of scientific determinism that developed in the mid-nineteenth century. Whereas Newton indicated a determinate mechanism at work in natural phenomena, the scientific determinists extended that principle to the human mind as well. The biological determinism of Darwin as outlined in *The Origin of Species* followed in step with the notion that man physically and mentally is a determinate creature. This of course overturned the basic elements of religious doctrine, and with it alienated the theological soul from the scientific personality.

With the widespread effluxion of scientific determinism, the spirit of atheistic materialism rose to a point of ascendency over theistic idealism and it maintained a predominant position throughout the twentieth century. On its heels came a rapid decline in the popular regard for religious values and theological principles. The irony of it all is that science does not intentionally assume a posture of antipathy toward religion. Officially, science offers absolutely nought by way of comment either about specific religious doctrine or general theological principles.

Religion and theology try to understand the universe and man's place in it. Man is unique in that he is blessed with a conscious awareness far more subtle and discerning than any other species of his acquaintance. Through the amazing alertness of his ego, man is able to puzzle about the nature of his environment, the universe, and more strikingly, about his own peculiar identity. Among these various facets of enigmatic postulation, science engages in only one area of speculation, and this concerns the physical being within the physical universe. In this restricted discipline, science itself is fragmented into an assortment of separate and independent subjects. A biologist need know nothing of astronomy, and a chemist may be ignorant of higher mathematics. Such lack of knowledge is hardly shameful. Indeed, it has been suggested that the very strength of science is due to the distinct fragmentation of its numerous branches.

By the same token, the scientist is hardly expected to have mastered the advanced concepts of knowledge outside the domain of science, such as philosophy or theology. It is not a mark of embarrassment if the scientist acknowledges illiteracy in these areas. More to his credit would be his exhibition of readiness to turn to the specialists in the fields of philosophy and theology for instruction and guidance.

Some critics consider that the passionate desideration of science for increased knowledge is an indication that science will never be content until it has mastered the secrets of an entire unified universe. Others protest and disclaim any such ambition, since they assume that there can never be a science of the universe as a whole. In fact, they conclude that the more science discovers, the more is it likely that new and different subjects may develop under its broad canopy; and as a consequence, the individual subject areas will expand in further diffusion and separation.

66

An ultimate ambition on the part of science is to arrive at a meaningful station of overall knowledge from whence man may derive a satisfactory comprehension of his position, purpose, and destiny in life. In line with this design, science may only generate partial conclusions or theories. Yet from such a stalwart base of information, the creative imagination of the human intellect may produce concepts of extraordinary depth and discernment. This adventure, of course, must lure man into the intriguing dominion of the abstract. It is quite possible that once immured within the world of philosophical and theological speculation, man may find it necessary to make a daring leap of faith.

Such an opportunity should not perturb the twentieth-century scientist, because within the bounds of his own disciplines a leap of faith was assumed on several occasions and the results were intuitively rewarding. Consider the excitement generated by Einstein's theory of relativity in the early years of the twentieth century. Its implications shook the very pillars of scientific determinism. At the outset, there was no way to test its validity. The scientists who accepted its credibility, could do so only as an act of faith. On the other hand, its recognition would indicate a challenge to the undisputed principle of scientific determinism. Courageous scientists, in those early inconclusive days, boldly took that leap of faith, and by doing so, they incautiously placed their scientific reputations on the line. The rest of the tale is now common knowledge. Einstein was vindicated and proved correct. A revolution ensued in the world of physics and related sciences. The behavior of the light of stars acted in accordance with the remarkable conclusions of an ingenious intellect, and those who indulged in the early leap of faith were rewarded by the ultimate triumph of their convictions.

If scientists find it appropriate to boldly forge new dimensions by virtue of imaginative assumptions within the ranks of their own discipline, then it may not be amiss to suggest their reliance upon the intuitive powers of fertile philosophic or theological minds, who may in somewhat parallel fashion call for a leap of faith. Perhaps at this delicate point in scientific history when the ground roots of scientific determinism have been overturned by recent exposures arising through relativity and again through indeterminacy, as conceived in the particle nature of intraatomic physics, the time may be at hand when courageous figures on the scientific scene may step into a new role of dynamic leadership by affirming that a theistic position is a proper and honorable one for principled men of science to adopt.

3

SOME PRELIMINARY COMMENTS ON "HOW" AND "WHY"

The area of "how" and "why" has received energetic attention in recent philosophical speculation. Specifically, does the "why" of *existence* intrigue the human mind. Will we ever be able to comprehend why a universe came into being, and why on a nondescript planet, orbiting around an insignificant star, in an inconsequential galaxy, an *elan vital* should produce a thriving habitat climaxed by an intellectual anthropos? This is a mind-staggering query. Professor Lovell indicates in his superb cosmological volume that science, curtailed as it is to an investigation of the "how," should acknowledge its limitations when it arrives at the point of the metaphysical "why."[2]

Let us endeavor to succinctly clarify what is implied by the notions "how" and "why." Scientific research is concerned with the "how" of things. Philosophical metaphysics and theology are absorbed in the "why" of the

---

2. A. C. B. Lovell, *The Individual and the Universe,* The New American Library, A Mentor Book, (New York, 1961). By arrangement with Harper & Brothers, pp. 124–26.

universe. These are two very exclusive and restricted areas of investigation. The scientist need not be concerned with an inquiry into the "why," and the metaphysical philosopher or theologian should find it unnecessary to speculate about the "how" (although "how" knowledge would greatly elevate his level of sophistication in pursuit of the "why"). This clear-cut separation of research fields and interests gives one pause to wonder about the long apparent feud that raged between science and religion in past generations. There really is no conflict of interest between the two. An individual may be an ardent scientist and still adhere to theological principles. Similarly, a theologically imbued person should find it possible to pursue a scientific career without surrendering enlightened religious scruples and practices.

In fact, it becomes rather simple to differentiate between the two categories. When the scientist, tempted by the fascinating spectre of the "why," enters into its analysis, he is no longer the scientist, but a metaphysician. In co-ordinate fashion, when a theological soul delves into the "how," he cannot assume to make such an analysis as a theologian, but rather he must commit himself to scientific principles of research.

Etienne Gilson endorses the logical differentiation between "how" and "why." He admits to the staggering dimension of the problem of *Being*, and he goes on to explain that the "why" leads to ever deeper inquiries. "Why are there organized beings?" is not the ultimate question. One may use the terminology of Leibniz and ask further: "Why is there something rather than nothing?" There are scientists who consider this problem as meaningless, and from the point of view of the scientist, this may very well be so. Since science does not entertain a "why" question, the "why" makes no sense. Science speculates in the area of the "how"; therefore, the area of the "why" is excluded by

69

science "precisely because it cannot even ask the question." [3]

Now, although the scientist, in principle, is bound by the limitations of the "how," the metaphysical investigator need not restrict himself to an exclusive "why." The latter may also be cognizant of the emergent knowledge produced by scientific "how" investigators, since it is through such information that he may better comprehend the manner of structuring the "why." An aspect of the Divine Personality is concealed in the "how." This is not to say that given scientific knowledge alone, the "why" becomes clarified. If so, then science would potentially be able to arrive at the answers to all of life's problems, because this is what the "why" entails.

Incarcerated as it is in the "how," Gilson is quite emphatic in his declaration that science can never accommodate the "why," because it lies outside of the sphere of physics, chemistry, and biology. He states that the most exhaustive scientific knowledge would not be able to explain why the world is made up of its component things. In our world of change, science can teach us "how" such change occurs. What science "cannot teach us is why this world, taken together with its laws, its order, and its intelligibility, is, or exists." [4]

Gilson forcefully reiterates this point in his reference to the classic work *The Mysterious Universe*, by Sir James Jeans. Gilson views the title itself as further revealing the impotency of science in its confrontation with the "why" of things. On what basis does the scientist allude to this universe as a "mysterious universe"? Is it because the more science advances, the more difficult are the problems that it confronts? But, states Gilson, "the unknown is not necessarily a mystery . . . because it is at least knowable, even

3. Etienne Gilson, *God and Philosophy:* (New Haven: Yale University Press, 1969), p. 139.
4. Ibid., p. 71.

though we do not know it yet."[5] The reason some scientists imagine the universe to be mysterious is because they mistake existential or "metaphysical questions for scientific ones," and "they ask science to answer them. Naturally, they get no answers."[6] In their puzzlement, scientists assume that the universe is mysterious, concludes Gilson.

Modern scientists of strong positivist and even agnostic persuasion readily concede that science at best can only explain "how," but not "why." Science, for example, through medicine, can explain *how* the body functions in the digestive process, but it can never adequately explain *why* it is so. The laws of gravitation can be understood in the *how* sense, but not *why*. Why bodies, elements, or particles should influence each other remains unanswered. The "why" remains a mystery.

Great care should be taken lest a scientist mistake a "how" question for a "why" question. The good scientist should not be carried away by semantic confusion and conclude that because, grammatically, a "why" may appear to be appropriate in certain phrases or sentence structures, it thereby indicates that a "why" question is expressed. Moreover, the scientist should not assume that when the "why" and "how" are interchangeable in certain contexts, a "why" interrogation has been sponsored. It must be made clear that when "why" and "how" are given the same meaning, then the "why" must be recognized for what it truly represents, which is nothing more than a semantically distorted "how." The sound and appearance may be that of "why," but its actual essence and intent is that of "how."

It is not possible for the scientist to ask anything but "how" so long as he is robed in his laboratory frock and remains incarcerated in his research chambers. "Why" can

---

5. Ibid., p. 123.
6. Ibid.

71

only be spoken by Philosophical Man or Theological Man, but never by the scientist as such. We shall endeavor to explore this matter more systematically at a later point in this work.

Even in connection with the "why" query we must be alert to the possibility of semantic confusion. Our concern is only with the cosmic "why" and not with a personal "why." The *cosmic* "why" relates to problems of the universe commencing with the query, "Why—a universe?" The personal "why" is not really a "why" question at all; rather is it an expression of complaint, very much like the grievance that Job uttered when he perceived the undue suffering of the righteous. Such a "why" is a moral determination, based largely on apparent inconsistencies between man's concept of divine justice and its failure to dominate the environment and society. Man assumes that God is duty-bound to aid, protect, and reward the righteous. "Why do the good suffer?" is not really a question within the periphery of the cosmic "why," despite the fact that it may eventually develop as a consequence of the cosmic "why."

One must also be cautioned against identifying as a "why" query what I choose to call the *absurd* "why." This interrogation follows Heidegger's hypothesis that the very existence of God, Himself, must be questioned—in other words, "Why—God? Why not—nothing?" The absurd "why" does not qualify as a "why" question for several reasons. First of all, it is logically self-defeating. The human intellect rejects the notion that nothing produces something. (This is not to be confused with the *ex nihilo* concept, wherein God, a very definite Supreme Being—i.e., a "Divine Something"—is conceded the omnipotence to produce a universe from nought.) Secondly, by its very definition, we mortals have chosen to identify the term *God* as designating the concept of an unsponsored or uncaused existent being. Heidegger's murmur reflects the classic

72

plight of the atheist, who, by virtue of his illogical position at the climax of a cosmological inquiry, attempts to thwart the incongruity of his assumptions by feigning a belligerent posture. Thus, if at the end of a cosmological safari, when the atheist must explain the appearance of the first particle of energy, he would audaciously conclude: "Nothing!," there is no doubt but that he would then be roundly ridiculed for expounding a preposterous notion. As a subtle strategem, therefore, the atheist primarily adopts an offensive stance, which momentarily relieves him from admitting to an embarrassing, ludicrous position; and so he precipitously advances the challenge: "Why was there God at the beginning rather than nothing?" This is merely an expression of the absurd "why." It is hardly a precocious query.

Rather often, four-year-olds express the same thought when they seek an explanation as to their existence.

"Who made me?" asks the little tot.

"God made you, dear," replies the loving mother tactfully.

"And who made Mommy?" he pursues.

"Why, God made Mommy, too," explains the parent.

"Did God make Daddy?" inquires the youngster.

"Why, certainly," responds the mother.

There is a pregnant pause. The little one digests these surprising clarifications, and then he perspicaciously demands, "Mommy who made God?"

Quite clearly, the absurd "why" is childish.

Let me make this point clear. It is not my intention to belittle the intelligence of philosophers who stir up a great furor over the question of God's own supposed "beginning"; it is simply that the finite mind cannot be bent beyond a certain point. The term *God*, like the concepts "first-cause" or "unmoved mover," stagger the human intellect. A primary principle in conceiving the meaning of God is that we are discussing a Being outside the pale of time and space, concerning whom it is impossible to attribute either a beginning or an end. Whether we say God is eternal or timeless, what we are attempting to express through the frugal capacity of our finite intellect is that the superior nature of an infinite, omnipotent, and presumably omnipresent Being can never be fully comprehended or known by man. We may not treat the term *God* as if it were just another proper noun attached to some concrete personality of our acquaintance. A metaphysical or theological insight or awareness of God is a far cry from truly "knowing" God. By virtue of the very definition of the word *God* we can never inquire, "Who made God?"; therefore, such a query becomes self-contradictory and absurd. The question itself is nothing more than an indication of the deplorable deficiency of the human intellect that man must tolerate as a finite creature.

Science endeavors to find solutions to questions in the "how" category; but can answers be forthcoming for questions pertaining to "why"? At times, an inspired soul such as Etienne Gilson may attempt to supply an answer. "Why are there organized beings? Why is there something rather than nothing?" Gilson explains: Every existential energy and thing owes its existence to a pure "Act of existence." The supreme cause "has to be absolute existence." If it is agreed by the scientist that the cause for the existence of organisms comes from outside their physiochemical makeup, then the cause for their existence transcends the physical order. Such being the case, the cause for the existence of the organism, whether it be as minute as a proton, neutron, electron, or quark, must, of necessity, be transphysical, or better yet, metaphysical. Thus, if nothing in the elements themselves accounts for their determined design, "the presence of design in a chaos of elements entails just as necessarily a *creation* as the very existence of the elements."[7]

The absolute cause is self-sufficient, and if it so wills, it creates in accordance with its free will. Moreover, since we witness as the products of its creation, being accompanied by manifest order, it itself must possess the only principle of order we know, namely, thought or intellect. Once an absolute, self-subsisting, and knowing cause is determined, we may no longer consider its identity as an "It," but rather a "He." "In short," states Gilson, "the first cause is the One in whom the cause of both nature and history coincide, a philosophical God who can also be the God of religion."[8]

Gilson comes to a noble conclusion, but the latter part of his quoted phrase is somewhat disconcerting. I cannot agree that the philosopher's God is also the theologian's God; nor is it the God of popular religion. Since we explore this point more fully later on, let us leave the topic in abeyance so that it may be taken up in a more appropriate perspective when we discuss the variations entailed in ego personality identities.

4

SCIENCE AT AN IMPASSE

Science performs best when it deals with inanimate matter. It can examine with precision the size, weight, measurements, chemical makeup, and relative position of an object. It is also possible to confront living organisms on this basis, yet the temptation abounds, particularly in an analysis of living organisms, to put forth the question "why." What is the purpose of the physical and chemical developments that abound in organic species?

7. Ibid., p. 140.
8. Ibid., p. 141.

It is the more predominant attitude among scientists today to assume that life has no special purpose. The broad influence of Darwinism has to a great extent helped to propagate such a point of view. Along these lines a whole body of current existentialist thinking has made the acknowledgment of purposeless existence its central focal point for developing a philosophical analysis of man. If life is meaningless, of what value are moral commitments? If life has no purpose, why should the individual assume any ethical responsibilities? If, in accordance with scientific thinking, the philosopher of an atheistic existentialist bent assumes that there is no "why" category, then life with all its anguish must be a hopeless and futile endeavor.

This nihilistic concept of the absurdity of life stems to some extent from science's aloofness toward the problem of existential purpose. Surprisingly, however, we find a contradiction in the very element of scientific revelation itself. Consider, for example, Darwin's theory of natural selection. When it is scrutinized critically, it leaves more to the imagination than to observable facts—and science, by all means, thrives on observable facts.

Darwinism as a doctrine taxes the credulity of the human brain. Indeed, it calls for an act of dedicated faith. We are asked to accept that the whole evolution of living organisms developed without any plan, design, or purpose beforehand. The biological status of our planet arrived at its present condition by the two-fold process of random variations and the struggle for existence. What is the marked weakness in this assumption, which causes the logical mind to recoil in disbelief? Very simply, the Darwinian theory fails to clarify the most startling aspect of the whole evolutionary process, namely, that nature has always proceeded from a lower to a higher species.

Let us analyze the problem further. If life holds no purpose, why should nature have produced any living organism in the first place? What impels this drive on the part of matter to emerge as a throbbing, moving, independent animus? In addition, accepting the fact of life, why should living organisms be so intent on perpetuating their own kind? Moreover, if it were merely necessary for a purely accidental surge of life to make its appearance and to survive as its own designated species, why should it have made the leap, again and again, to graduate from species unto higher species on a long, tortuous chain of procession upwards, since there was no purpose for any organic existence at the outset? Finally, if mere survival is the whole essence of its accidental appearance, it would seem to be the epitome of logic for some very elementary type of organism to make its accidental debut, and that should be the sum substance, and total end to the story of organic existence. Why this incontrovertible drive to climb the ladder of development to a pinnacle of intelligent life topped by man? Parenthetically, there seems to be no valid reason for even a rudimentary organism to have evolved as a living entity, if all that was desired was a prolonged occupation of space in the form of matter. The inanimate rocks and inert chemicals have existed intact for billions of years without resorting to the bothersome process of care, feeding, nurturing, reproducing, and growing. From the point of view of a purposeless nature, the simple boulder outrivals any organism, if but an existent object in time and space is all that was required. The Darwinian notion is a hard pill to absorb by all standards of logical permutations and rational thinking.

There are many unsolved problems on the distant cosmological horizon. Professor Lovell traces a number of these challenging enigmas in his superb work, *The Individual and the Universe*,[9] where he considers such mind-

9. A. C. B. Lovell, *The Individual and the Universe*, The New American Library, A Mentor Book, (New York, 1961). By arrangement with Harper & Brothers.

baffling questions as the origin of the universe, the beginning of time, and the birth of our solar system. For example, despite the marvelous and extensive apparatus that is presently available for the investigation of outer space, one regrets that the astronomer must be contained by the limited scope of his instruments. When the astrophysicist has reached the distance of a few thousand million light years, the light and radio waves become so weak that it is no longer possible to observe what lies beyond. Given the human mind's penchant for curiosity, who can be blamed for asking, "What lies beyond that last frontier of contact?"

Another exciting feature of astronomical research is the fact that the farther out the astronomer peers into space, the farther back in time does he reach. When he regards the sun, he sees it at the position it occupied eight minutes earlier; the nearest star is sighted as it was four years prior; and the more distant extragalactic waves tell us what they were like millions of years heretofore.

It is now generally agreed that the question of a magnetic field in the cosmos and how it came to be has an important bearing on our view of the entire cosmological process. In order to trace the origin of the magnetic field back to its source, it is necessary to commence the sleuthing with our nearest star—the sun. It seems that the sun's magnetic field was not a spontaneous occurrence, rather was it a derivative of the gas cloud from which it condensed about 4,500 million years ago. How the gas cloud came to possess a magnetic field is a difficult problem; however, astronomers are generally of the opinion that the field came along as part of the other elements that were imparted from its source, namely our galaxy, the Milky Way. Proceeding one step farther, it is assumed that our galaxy, like all other galaxies, received a magnetic field from the basic root of the universe; and finally, it is concluded that magnetic fields are found throughout the wide expanse of nature.

By pushing the ultimate source of a magnetic field back to the very origin of the universe, it becomes a very important segment of cosmological investigation. Why does the universe itself possess a magnetic field? The answer denotes a clue as to how the universe arrived at its present condition.

There are two major theories that currently appeal to the scientific mind. They are denoted by somewhat crass designations: the "big bang" theory and the "steady state" theory. Since the big bang theory accepts the premise that the universe had a beginning in time, it makes possible an explanation of numerous problems in astronomy and physics, including such an item as the magnetic field. By postulating the fact that the universe itself was endowed at the outset with a strong magnetic field, this enigma is handily solved for all components of the universe.

The steady state theory, on the other hand, cannot explain the origin of the magnetic field that permeates the universe. Furthermore, it exhibits weaknesses in other respects. The steady state theory fails to clarify the radiation of a microwave background, and how such a smooth setting of radio waves, which uniformly appear throughout the heavens, came into being in the first place. Although the source of a microwave background is currently an unanswered problem in astronomy, at least the big bang proponents can offer a solution by theorizing that it stems from the early formation of the universe. The steady state adherents, who consider that the universe had no origin, are left without an explanation.

It is not my intention to dwell in detail upon the precise aspects of recent cosmological thinking. The reader who is so inclined may discover a mine of information on the subject in expertly written accounts by some of the world's leading authorities. Personally, I am partial to the thoughtful analysis and clarification set forth by Professor Lovell, particularly in his masterful presentation, *The Individual and the Universe.*[10]

Professor Lovell rightly considers the question of cosmological origins as the greatest challenge ever posed to the human intellect. Thanks to the technological marvels of modern instrumentation, stellar phenomena some two thousand million light years away are currently brought within the range of human examination. This means that man, today, can inspect areas of the universe as they appeared two thousand million years ago. With such emergent facts at hand, it may even be possible to infer the nature and extent of the universe beyond the border of human observation.

As Professor Lovell sees it, there are three basic aspects to the cosmological problem and these are involved in the following inquiries: (1) Will it be possible to penetrate farther into the universe, beyond the present level of contact? (2) How do present observations agree with cosmological theories? (3) How recognizable is the need for a transition from physics to metaphysics, and from astronomy to theology, when science exhausts its technical exposition of the subjects?

For any discussion of the first two problems, we shall necessarily rely upon the judgment and conclusions of experts in the field. In the third area of inquiry, however, we shall endeavor to present some new avenues of thought for serious consideration and reflection.

Impressive, indeed, is the information that over the entire range of the observable universe that extends into vast reaches of space and time, and that enfolds several hundred million galaxies of stars, a high degree of uniformity is constantly found. The constancy of uniformity abounding

---

10. Some of the astronomical information in this section is basically drawn from chapters 5 and 6 of Professor Lovell's work, *The Individual and the Universe.*

in the universe is one of the most significant factors for concluding a cosmological origin in space and time at the behest of a supreme intellect. Uniformity arising as an accident may occur spasmodically on a one-time basis in a remote corner of the universe, and hardly ever repeat itself again. This would be in accordance with the most favorable odds offered by the mathematician. However, when uniformity is found to prevail everywhere, with little exception to its common appearance, then we cannot but assume that law is an incontrovertible essence of nature, and that it was not a rare, accidental surge expressed by turbulent particles of matter, but that it was foreordained, implanted, and univocal at the outset. At this point it is not even necessary to cross into the area of advanced metaphysical speculation, because in accordance with the most elementary and basic foundation of logical apprehension, the appearance of spontaneous law seems highly unlikely unless such law is propounded by a lawmaker. It is impossible by any stretch of the imagination to conclude otherwise.

Yet, largely prompted by the impact of Darwinian determinism during the past century, science has been content to adhere to the notion of a godless universe. Matter was conceived as its own sponsor, its own designer, its own producer, and its own destiny maker. In this era of advanced computer technology the tenor of such doggedly persistent prejudice is gradually diminishing in favor of a broader spectrum of tolerance for a logically inferred programmer behind the cosmic scheme. This new spirit of acknowledgment that a divine power exists behind nature's facade is currently gaining a wider acceptance, although it is still meeting with rugged resistance in certain scientific circles.

Despite the fact that scientific interest in the area of metaphysical and theological assumptions has been gathering momentum, the overall mood is still one of caution and reserve. Consider, as an example, the following genial banter that recently took place in a biochemistry class in Cornell's Graduate School of Medical Sciences. The professor was clarifying the role of RNA in the living cell, and he traced its development from the programmed molecules of DNA. The class was comprised of young students who grew up in the era of computer technology, and therefore, were quite conversant with the necessary role of the programmer in formulating a computerized program. The mention of "programmed molecules" prompted an alert inquiry from one of the students.

"If the RNA and DNA structures are programmed to perform and produce in accordance with their designated patterns," he asked, "from whence come the programs themselves?"

The professor responded with a whimsical smile, "Why from the Generator of Diversity, of course."

The puzzled student quizzically exclaimed, "The Generator of Diversity? Who is that?"

"Well, I guess He is better known by His initials," came the genial mentor's rejoinder.

Mention of God arises in the halls of science, but at this stage, His identity is passed along in furtive whispers.

"Who is responsible for it all?" the curious mind inquires.

"Don't mention that I said so," replies the uneasy sage of science, "but the answer appears to be God."

The astronomer is perhaps more conscious of the metaphysical and theological problems involved in a comprehension of nature than scientists of other disciplines, because in scanning the vast expanse of space, he is constantly confronted with the awesome majesty of the universe. By comparison, the diminutive stature of man encourages the assumption of an humble posture in the course of research. The astronomer realizes that his view of the cosmos, while overwhelmingly mind-staggering by virtue of its enormity, may actually only embrace but a small segment of the true extent of the world of matter. Furthermore, it seems certain that due to the excessive recession of the galaxies, it will never be possible to substantially exceed the present observable limits of the universe. We may never hope to secure any information about the outer regions of space where receding galaxies travel with the speed of light.

With the application of radio telescopes the farthest reach of the horizon may yet be expanded by a limited distance of some several thousand million light years, but then, man will have reached the zenith of his potential sightings. It is a point of cosmological impasse for the astronomer. What lies beyond will forever be sealed outside the limits of human observation.

When the astronomer has exhausted his physical fount of knowledge, he must resort to a consideration of cosmological theory as the next step for advancement. As Professor Lovell explains, most astronomers today favor the evolutionary models of the universe developed in accordance with Einstein's general theory of relativity. Einstein was unable to produce any solution in his equations that would describe a static universe. Eventually observable facts proved the correctness of Einstein's assumptions. The universe was found to be expanding rather than static.

Three hypotheses present themselves as possible interpretations of a nonstatic universe: (1) a universe that commenced from a point of origin at a specific moment in finite time, and expands to infinite proportions in infinite time; (2) a universe with a specific radius at the early beginning of time, which eventually continues in infinite expansion and infinite time; (3) a universe "which expands from zero radius to a certain maximum and then collapses to zero again" [11] after which the whole process is repeated interminably.

---

11. Ibid., p. 96.

An evolutionary model that gained popular favor by contemporary cosmologists found its source in a solution presented by Abbe Lemaitre in 1927 and expanded by Sir Arthur Eddington. Eddington held that at the outset the universe was comprised of a uniform collection of protons and electrons. The radius of this gaseous primeval entity was about a thousand million light years. Some undetermined irregularity occurred, and the universe was prompted into a mode of expansion. Once the process of expansion commenced, the evolutionary development followed, and the universe grew to its present size, approximately five times that of the initial static proton-neutron gas cloud.

Eddington's theory in itself suffered from several shortcomings. First of all, his time scale for the evolutionary development was too brief. Secondly, he failed to explain why the primeval gas was disturbed to pursue a pattern of expansion rather than contraction. Thirdly, what brought about the fortuitous original conditions of a primeval gaseous state in the first place? Furthermore, to assume that it was only by chance that the initial disturbances were set in motion hardly assuages the human sensitivity to the notion of purpose, particularly when an entire cohesively constituted universe exhibiting a high degree of uniformity everywhere is considered to be the offspring of an ephemeral chance occurrence thousands of million years ago.

In the case of Eddington's theory, the "how" was never clearly satisfied; the "why" was clouded; and the "who" was veiled. Sir James Jeans, sensing the predicament, offered to clarify the latter dilemma by postulating the finger of God as agitating the original ether. It seems that Jeans was somewhat chastised by his colleagues, since they felt that a scientist should not become too familiar with the theological "who." Yet, even by scientific standards, his fellow astronomers could have been less harsh on Jeans. There is a question that every honest cosmic researcher must eventually confront, and that is "how" did the primeval gas originate at the outset? Science is evidently quite silent when it faces this problem. Logically, it appears, in Professor Lovell's words, "that the theory requires the exercise of yet another divine act at some indeterminate time before the occurrence which set off the gas on its career of condensation and expansion." [12]

Eddington's theory has been superseded by more sophisticated explanations based upon recent advancements in astronomical research. The influence of Lemaitre is still apparent in a now popular model of cosmic evolution that is structured in accordance with Einstein's general relativity concepts. The universe is considered to have originated at a finite moment in the past and is expanding to infinite size in an infinite future.

By theoretically reversing the cosmic process of expansion, we return to a time some eight or nine thousand million years ago when all the galaxies were possibly very close together. Moving farther back in time, a situation develops that indicates that the source of the entire universe was a small, dense conglomerate, commonly called the primeval atom. The concentration of density within the primeval atom was inconceivably high, in the neighborhood of a hundred million tons per cubic centimeter.

In some manner, an initial momentum scattered the material of the primeval atom. After thousands of millions of years, an Einsteinian universe developed. Its size was about a thousand million light years. It maintained a static condition of balance between the attraction of gravitational forces and cosmical repulsion for aeons. During this period, "the great clusters of galaxies began to form primeval material. Then the conditions of near equilibrium were again upset," [13] and a second expansion commenced as the

12. Ibid., p. 99.
13. Ibid., p. 103.

forces of cosmic repulsion overcame the forces of gravitational attraction. After nine thousand million years, the universe arrived at its present state of appearance.

Science feels confident that it can deal with the period commencing at a time several thousand million years ago, when the universe was in a state of original gaseous clouds. That inconclusive era was hardly the point of onset of either time or space. It was merely a condition that enabled the clusters of galaxies to form and develop. Such a segment of galactic evolution can be dealt with through mathematical analysis, and its expansion can be described in terms of present-day physics.

An impasse arises when an attempt is made to pierce this boundary and to construct the cosmic scene of earlier periods. It is not known how much farther back in time the primeval atom made its debut. A date for this event cannot even be approximated, since it is impossible to surmise how much time elapsed during the galactic incubation period, when the gaseous clouds were in a state of gestation with their cosmic offspring. If the astronomer must offer a figure, he would suggest that the primeval atom gave birth to the universe through an act of tremendous explosion or disintegration between twenty thousand million and sixty thousand million years ago. On this time chart, the formation of the galaxies about nine thousand million years ago, leading to their present state of expansion, can be considered as a relatively recent event in cosmic history.

Science believes that it may speak with authority when it encounters the "how" of cosmic birth and development. It feels that it can explain the general evolution of the universe commencing with the primeval atom's eruption. However, any discussion about the nature of the primeval atom itself—how it came into being; when it emerged; of what matter it was constituted; or what made it spew forth

86

a universe—is beyond all powers of investigation by the discipline of science. Here is an impasse that science could never conceivably conquer, because there will never be available to the scientist the opportunity to either observe or deduce the original conditions that prevailed at that time. In fact, it relates to a situation that science may consider as outside the realm of time, since the whole universe including space was contained within the primeval atom. If an analysis is to be attempted, then one must vacate the domain of science and astrophysics, and enter into the expository realm of metaphysics and theology. The speculative edifice for a consistent comprehension of the cosmic origin must emanate from the pen of the philosopher, and if he constructs a logical path leading to the emanation of space-time, the mathematician and physicist may handily pick up the trail and continue the investigation.

There is another alternative as Professor Lovell so aptly puts it:

> Or one can simply refuse to discuss the question. If we wish to be materialistic, then we adopt the same attitude of mind as the materialist adopts in more common situations. The materialist will begin in the present case at the initiation of space-time when the primeval atom disintegrated. That quite simply evades the problem.[14]

What would be the nature of metaphysical speculation concerning the cosmic origins that a scientist may find somewhat to his taste? Professor Lovell offers a pattern of investigation in that direction.

To begin with, when we consider the primeval atom and the struggles it postulates, such as in the area of time and space, we may very likely reiterate Professor Lovell's reactions to such a challenge: "I feel as though I've suddenly driven into a great fog barrier where the familiar

---

14. Ibid., p. 106.

world has disappeared." [15] Anyone who has ever experienced an English fog can well appreciate Professor Lovell's reaction.

A basic problem in conceiving the origin of the universe is to ascertain the change from a condition of indeterminacy to a state of determinacy, following the commencement of space and time, when the laws of physics in a macrocosmos are applicable. Upon arriving at this point of departure. the scientist may feel a little more at home. Indeterminacy calls to mind the quantum theory of physics where the behavior of particular atoms is studied, in contrast to the determinacy that is possible in situations pertaining to large numbers of atoms. In the latter case, it becomes possible to formulate predictable behavior and hypothesize results. In the former instance, it is not a simple matter to pinpoint the attitude or consequences of individual particle behavior. The principle of uncertainty explains why the particular atom is indeterminate, since the very endeavor toward making it the object of investigation disturbs its attitude of behavior.

By applying the quantum theory to the cosmological problem, it becomes possible to at least engage in some sort of analysis as to the beginning of space and time. The primeval atom may relate in parallel fashion to the individual atom of the quantum theory. The primeval atom was a unique phenomenon unattached to any of the physical laws of the universe with which we are familiar. Perhaps it operated in accordance with some unknown primeval law, but that is not for us to say. Like the individual particle in the uncertainty principle of modern physics, the determination of its procedure is unpredictable and analytically impossible because its behavior could have been totally chaotic at the outset.

Once the primeval atom disintegrated and produced a

---

15. Ibid., p. 108.

state of multiplicity, it was possible for the universe to assume a determinate identity in a macroscopic sense. Such a situation could be correlated to the state of determinacy that operates when large numbers of atoms are involved.

Professor Lovell identifies the beginning of space and time as arising with the condition of multiplicity, and he adds, "but the beginning itself is quite inaccessible. In fact, in the beginning the entire universe of the primeval atom was effectively a single quantum unit in the same sense that only one of the future innumerable potential states existed." [16]

By identifying the primeval atom as an original cosmological quantum, it offers hope to future researchers, that in the event progress may be made in comprehending the nature of the individual atom, it may become possible to draw a similar parallel to the condition of the primeval atom, and thereby shed further light on that primordial, dense conglomerate of matter.

Professor Lovell proceeds to outline a suggested method of cosmic development from the primeval atom. One can only postulate aeons as the time element involved in the transformation of the intense energy within the primeval atom into gaseous clouds of hydrogen. High pressures and temperatures also explain the further development of the hydrogen clouds into other elements at the time when the stars commenced to take shape. Says Professor Lovell:

> If pressed to describe this primeval atom in conventional terms one would, I think, refer to a gigantic neutron. By radioactive decay this neutron suffered a tremendous explosion. Protons, electrons, alpha particles, and other fundamental particles emerged from it at great velocity and continued to fill all space nearly uniformly as this basic material expanded for many thousands of millions of years until the clusters of galaxies began to form.[17]

---

16. Ibid., p. 109.
17. Ibid., pp. 110–11.

Professor Lovell offers an interesting cosmic explanation of the universe in terms of astrophysical dynamics and scientifically dominated metaphysical speculation. We must of course realize that the essence of the primeval atom itself remains unexplained. It is a point beyond which science can proceed no further. To satisfy this quandary another route must be taken, and Professor Lovell cites an unidentified leading theologian who abides by the evolutionary theory of the universe, but patently pursues the matter to a logically ultimate conclusion by asserting that "the creation of the primeval atom was a divine act outside the limits of scientific knowledge and indeed scientific investigation." [18] Simply paraphrased what this means is that in the beginning God created the primeval atom. The rest is cosmic history.

The steady state theory appears as an alternative to the above and seems to satisfy those who reject God, since they assume a strictly materialistic attitude toward the emergence of the primeval atom. For them, the creation of matter is a continuous affair, although from a cosmic point of view the universe continues to maintain a steady state of relationships between stellar phenomena. As the galaxies recede from each other at tremendous speeds, and eventually disappear from view, they are replaced by other newly formed galaxies, so that the universe always continues to exhibit the same constitution of relationships between stellar phenomena. Actually, the individual galaxies will have come and gone, but the average spatial density throughout the universe continues undiminished because new matter and new galaxies are always in the process of formation throughout the universe, leaving the universe unchanged through all space and time.

The steady state theory leaves much to be desired from any frame of reference. First of all, it removes any possi-

_____
18. Ibid., p. 112.

bility of a beginning for this material universe that we daily behold. It is considered never to have had an origin either in time or in space. The universe is described as eternal, having existed from an infinite past, to the present, and it will continue on into an infinite future.

Considering that we are dealing with a universe of physical matter, the steady state theory leads to an unsavory materialistic pantheism. All that is identifiable is an immense infinite universe existing in a state of perpetual motion. It is difficult to conceive of a universe that is at the same time concrete, infinite, and timeless, despite the attempted explanation that there is an endless expansion and a replacement by new galactic formations. Can there be an endless expansion of physical substance in an infinitude that is already filled with stellar phenomena? Our finite minds meet with a great deal of trouble in endeavoring to comprehend such a postulate. To what empty point in space are the galaxies receding, if all of infinite space contains distinctly categorical matter?

Our finite intellect must further inquire: If such a universe is infinite to begin with, and appears as it does now, namely, well-stocked with stellar phenomena, where and how can there be room for further expansion? There seems to be a snag here in satisfying the concept of infinity. Moreover, once we deal with matter in terms of known laws of physics, we have eliminated the element of infinity. Matter and infinity are mutually exclusive. One cannot postulate a finite infinity, nor can the human mind accept an infinite finitude. It may be possible to consider space as infinite and devoid of matter, although this notion may now contradict some of the startling principles developed in the theory of relativity, specifically the concept that space curves in on itself, which negates a claim to infinitude. Besides, even the notion of infinite space is not in consonance with the idea of an infinite steady state universe. Finally, there remains unexplained the identity of the intellectual sponsor of observable physical law and order throughout the cosmos. We, therefore, must dismiss the steady state theory as logically inexpedient.

I am inclined to consider that some of the sponsors of the steady state theory are motivated more by personal prejudice than by objective scientific conviction. It is possible to suspect that their cosmological predilections are strongly influenced by preconceived notions of a narrow, self-serving atheism. Their conclusions may even derive from deep-seated psychological motivations, so that they may be entirely unaware of the subliminal stimuli that prompt them to make their choice. After all, scientists though they profess to be, they are only human, and it is not an easy matter to project laboratory objectivity into areas of personal partiality. Suffice it to commend the whole denomination of scientific researchers who sustain an atmospherically pure posture of honest, dedicated objectivity in their multifarious pursuits and investigations. Civilization owes them a debt of everlasting gratitude for their unswerving allegiance to the noble principles of scientific methodology, and no amount of respect and admiration could sufficiently compensate for their fervent zeal and diligence in their appointed tasks. However, there are moments when their very personal inclinations may prompt them to assume positions that would reflect opinions not necessarily based upon the rigid doctrine of scientific discipline. It is possible for even the best-intentioned scientist to succumb to deep-seated psychological forces and to express views that are not wholly the outgrowth of objective scientific evaluation.

This, I submit, is what essentially helps to foster a presumptuous perspective in the minds of many scientists, and further directs atheistic cosmologists to seek persistently godless theories in composing a script for the cosmic drama. In every theoretical blueprint of the universe, there is reflected the inner convictions of its all-too-human designer.

Not surprisingly, it will adhere to materialistic suppositions if it stems from one committed to a personal atheistic persuasion; and correspondingly, it will reflect idealistic views if it evolves from a soul who is theistically inclined. Seldom does there appear an individual who attempts to approach the cosmological problem with an open mind and an honestly objective perspective. To such a researcher belongs deserving tribute.

What prompts the choice of a materialistic direction on the part of some cosmologists? Again, if I am permitted to suggest a psychological motivation, I would submit that man is much more comfortable in a totally physical universe. A world of matter promises a greater source of security than an inconceivably void cosmic expanse. The attitude calls to mind an age-old inclination of mankind to indulge in idolatry. The recurring human susceptibility to idolatrous worship is psychologically indicative of man's attraction toward the concrete; and so the history of man is filled with page after page of sordid image-worship. Man's leaning toward the adoration of physically contrived deities has its roots in the inner subconscious bias, which identifies security with a corporeal environment.

The subconscious fear and rejection of a nondimensional habitat impels some theoreticians to consider as a basis for a cosmic design the more reassuring and familiar milieu of a physically populated infinitude. In the old days, such subconscious longings were assuaged by a concupiscence for concrete deities and sensuous idolatries. Lamentably, behind the facade of the postulated steady state theory, one suspects an inclination toward the old idolatrous dispositions. The physically exorbitant expanse of the steady staters falls in line with the modern-day impulse to adore material splendor and to indulge in sensuous hedonism above all else. Does not modern-day materialism replace the image worship of former civilizations?

No one can deny being under the spell of a self-imposed

subjugation to a deity of one form or another. An object of deification is found in every person's life. Even the so-called atheist admits to an "ultimate" in his life. All that is necessary is to properly identify that which is extolled as life's "ultimate," and behold, the supreme object of such glorification materializes as the target of deification. For many people, such an "ultimate" is simply God.

In the steady state universe, where only space and matter are predicated, and where physical law and order prevail, there is nothing left to identify along the whole infinite expanse other than space and physical matter. It therefore must be assumed that matter, somehow, intellectually postulated its own laws of operation. Matter, consequently, emerges as the "ultimate"; and since the "ultimate" is in essence an object of adulation, the deity of the steady staters is the infinite universe itself.

The notion of the universe as God, and God as the universe, is a very disagreeable complexion of pantheism. One of the difficulties with such a notion is that by definition God is conceived to be infinite. True, according to the steady state theory, the universe is infinite; but, it is also infinitely occupied with matter. Now, the concept of a supposedly infinite universe, composed of infinite stellar matter, is in reality a physically dimensional universe, and since an infinite finitude is a contradiction of terms, such a universe defies logical premise. Our intellect finds it impossible to comprehend an infinitude composed of finite galactic particles. Matter is subject to measurement, and therefore is confined. Matter in space permits the possibility of spatial dimensions. An infinite steady state universe, therefore, can be subject to measurement—but the thought of a measurable infinity is antithetical to the semantic designation of the term. It appears that the universe of the steady staters promotes more problems for the cosmic enigma than solutions.

Professor Lovell indicates, "The conflict between the steady state and evolutionary theories is of the very greatest significance to cosmology and to human thought." [19] Even if an attempt is made to identify God with the steady state proposal by assuming the creation process to be a divine act that is proceeding continuously, the theory is still troublesome. Under such an assumption, it must be kept in mind that the creation of matter is a continuous affair, and Professor Lovell advises that, given advanced astronomical instruments, it may become possible to probe the very areas in space where such creation is presumably going on, and this may tend to disrupt the further implementation of divine involvement.

The major issue in cosmology today, nevertheless, remains the question as to whether the universe is in a process of infinite renewal, or whether it was fundamentally created in its entirety aeons ago. The determination of this issue strongly affects philosophical and theological attitudes. Professor Lovell points out that from a strictly scientific frame of reference there is insufficient observational data to determine effectively which cosmological theory is the correct one. Moreso, the creation of primeval material will never be grasped by scientific discipline.

In dealing with the concept of creation, it is necessary to resort to metaphysical speculation; and whether philosophy, theology, and science can be welded into a harmonious balance is a matter of quite personal conviction. What is certain is that the committed scientist must remain within the bounds of "how" inquiry. The potential revelations accorded to the area of "why" must emerge through other disciplines, such as philosophy and theology. When the astronomer asks "whence," he poses an inquiry that must be directed far back into the remote expanse of time and

---

19. Ibid., p. 117.

space to the original warrants of the Supreme Creator. When the theologian speculates on "wherefore," he directs his sight toward an accessible Divine Personality in the here and now. But this topic is treated to a more precise analysis in the sections ahead.

5

ELEMENTS OF A PHILOSOPHY OF SCIENCE

Before we embark upon an existentialist cruise through some of the broad streams of intellectual enterprise, it seems appropriate to examine a few elements of contemporary scientific philosophy.

A philosophy of science may relate to functional aspects of the discipline and endeavor to analyze the significance of theory, method, or fact. Among the items that could come under discussion are such references as the nature of law and theory; the validity of the scientific method; patterns of scientific concepts; the manner of defining scientific ideas; and the semantic of logic. The philosophical enterprise in these areas of discussion may more likely appeal to the professional scientist. A nonscientist would probably find little interest in discussions limited to such highly technical, detailed, analytical subjects.

There is a second area of philosophical speculation relating to science. This would involve an analysis of conclusions as they may relate to the destiny of man and his society. Such an investigation could possibly attract the attention of a broader audience outside the perimeter of scientific research. By employing the results of scientific investigation, a philosophical edifice could be attempted, and its resultant structure would endeavor to clarify the nature of man and his universe. In this area of scientific philosophy, science would be used to help explain some of the problems normally associated with philosophy proper. To engage in this latter application of science, one would not

96

necessarily have to be a professional scientist. In fact, it may be advantageous to assign this component of philosophical extrapolation to the nonscientist for two reasons. First, the nonscientist by virtue of his noninvolvement in scientific pursuits on a professional level is very much free from the bias of a limited, particular branch of the discipline to which the scientist is usually attached. The nonscientist, therefore, could be more objective in an approach to the problems involved. Secondly, certain philosophical problems call for cohesive applications of several different branches of the scientific discipline. They are subsequently not limited to a single branch of the subject, and in such a case the nonscientist could be just as suited to handle information in various subject areas of science as the professional scientist, who although unquestionably proficient in his own division, may nevertheless be little acquainted with other sectors of the scientific realm. This was bound to occur when the widespread fragmentation of science necessitated narrow, individual specialization in a particular concentrated field of study.

Of course, in certain branches of science, one area overlaps into another, and this is particularly notable in phases of astronomy that embrace physics, or in geology, which depends upon chemistry and physics to explain some of its facts. It has even been suggested that several branches of science should relate to the domain of history, since they are based upon a historical development inherent in the growth of a species or the evolution of inorganic and organic matter. Since such information calls for historical recording, certain scientists would have to be capable historians as well.

Whether it be an intermingling of the various branches of science or a relationship to history, the major challenge in pursuing a philosophy of science arises when one considers the problems of cosmology. Among all the branches of science, it is probably astronomy that has the most significant relationship to cosmology. Therefore, we invariably turn to the learned astronomer when we seek an explanation pertaining to a rationale of the universe, since we expect him to hold a key of sorts to the riddle of the cosmos and its beginning in time and space.

Time and space are themselves problems of challenging enormity. Are they distinct entities or do they exist only in relation to other aspects of matter? In the *Timaeus*, Plato identifies space as a receptacle. If there were no space, matter could not exist, just as in parallel fashion it could be said that if there were no oceans, there could be no water creatures. The oceans perform a somewhat similar function for fish as does space for matter. Space, however, is more penetrating. Whereas water does not completely saturate the creatures of the sea, space does totally pervade matter.

Although it is not our intention to enter into an extensive discussion of the space-time enigma, a brief reference may help indicate some elements of the profound subject area that a philosophy of science must consider. The question of time is an intriguing challenge to the human intellect. Sense perceptions occur in an order that we identify as time. We recognize a general flow of time in the universe, from an unalterable past, through a decisive present, into an unformulated future. The flow appears to be unending; it cannot be stopped; nor can it be reversed.

Today, our measurement of time is identified by a metric system related to a uniform flow, which is matched to precise instruments. But how do we adjust our instruments? Here, once again, the astronomer emerges as a valuable authority. He explains that sidereal time, which measures the rotation of the earth in accordance with a reference to a fixed, distant star, is more reliable than solar time, which simply follows the earth's revolution around the sun. More uniform than sidereal time are measurements accruing through mathematical equations based upon the laws of mechanics. The astronomer guarantees their precision because they are derived from our observations of nature.

At this point we seem to have hopped onto a carousellike train of conjecture. The laws of mechanics lead us to an identification of uniform time; but we must first validate a process of uniform time in order to discern the laws of mechanics. It appears that we are caught up in the vortex of a vicious circle of reasoning.

Ultimately, we must rely upon the uniformity found in nature, either at the microcosmic or macrocosmic level, in order to identify and establish a uniformity of time. Philosophically speaking, such an admission by science is tantamount to a confession that there is inherent in nature elements of operation that not only appeal to a rational faculty such as our human intellect because of their precision and consistency, but moreso, because these elements of operation advertise the awesome creative intellect of an unseen sponsor. The scientist justifiably takes pride in his ability to produce instruments for the precise recording of time metrics, and his greatest achievement in this task is to make his system of time measurement match the stellar metric of space. Should not the scientist appreciate the fact that his dependency for establishing a definition of uniformity arises from the precise behavior patterns of astronomical bodies, and that these laws and definitions were operative prior to his appearance and continue to function with amazing accuracy throughout the universe?

Law is imbedded in nature, and given the appearance of a sagacious species, such law can be perceived by a rational intellect. Can such a rational intellect be induced to further perceive that a precise continuum of uniformity in matter, which makes possible confident predictability and provides a reliable yardstick for measurement, cannot be a product of its own making? Certainly, by all rational conclusions, it becomes necessary to search for a Supreme Intellect in the "great beyond."

Many other complications present themselves in an analysis of time. For example, the order of time itself becomes an engaging puzzle. Is it possible to identify the order of things and the order of events in their relational occurrence by endeavoring to adopt an objective standpoint? From a rational perspective it must be assumed that in the flow of a temporal sequence the cause necessarily precedes the effect. To validly identify the cause, it is essential to refer to some independent criterion, and once again we can only fall back upon the processes of nature. Nature exhibits a method of mixture, change, and expansion, which demonstrates a procedure from disorder to order. We may also note a change from order to disorder, again displayed in a relation of causality between physical events. The physicist recognizes a serial relationship in the order of things and thereby concludes the reality of an irreversible process.

The causal order in the universe explored by the human mind is accepted as a necessary time order, and it would be futile to attempt a cosmological investigation unless it is recognized as a basic postulate. Cosmology, therefore, at least from a human standpoint, calls for the acceptance of a time order as an undisputed premise.

Now a difficulty arises when an attempt is made to identify the Supreme Intellect, who must be presumed to be the sponsor of natural law and order, and to involve such a Supreme Intellect in a space-time continuum. Certainly, as the sponsor of a space-time continuum, the Supreme Intellect cannot be said to be subject to its laws or limitations. God cannot be bounded by matter. How, then, can man expect to relate to God, since it is impossible to identify a Supreme Being in any form of relation to the physically constituted universe of dimensions? God, therefore, not only remains boundless and timeless, but very much out of reach. How, indeed, can a finite intellect hope to authenticate a mode of communion or communication with a supremely transcendent Creator? Let us reserve these problems for later consideration.

One further comment should be cited in connection with time. From a simple categorical point of view, time is merely a dimension of matter. If it is conceded that matter exists in a three-dimensional physical system, then time appears as a fourth dimension for the further precise identification of an object. What this means, in effect, is that in a hypothetical continuum totally devoid of matter, there can never be an element of time. In other words, where there is no length, width, or breadth, there is no time. A timeless eventuality can be conceived (if it be permitted to so phrase it) prior to the emergence of a physical universe such as we now inhabit. Before the first particle came into existence, time was a nonentity—it had no meaning. Time did not commence until the simultaneous instant when the first charge of energy made its debut in the universe.

Of course, this leads to a plethora of objections and questions. Shall it be said that a vacuum is a timeless, spatial entity? But the observation and utilitarian application of a vacuum is an everyday occurrence in the laboratory, and a multitude of experimentation involving vacuums proceeds apace in what is definitely a time-based spectrum. This question is a veritable challenge because there is nothing we can more aptly conceive as being barren of physical matter than a vacuum. The answer is that there is a difference between a relational vacuum and a nonrelational vacuum; between a vacuum existing within bounded areas of space, and a vacuum beyond the sphere of space, and this statement, of course, is dependent upon our manner of defining the concept of space. A nonrelational vacuum, free of spatial entities, is timeless. Of such a vacuum it may be said that it lacks any identification of a past, present, or future. It is said that God is timeless. Does this mean that God is a vacuum? Hardly so! God resides (if we may use the term) outside of all dimensions. In truth, there is very little we can say about God, and perhaps all we can do at the mention of His identity is to open our mouths in awe-stricken astonishment and phonate a moan of absolute bewilderment. At best, we can offer only the vocalized symbol for His being, after which we have exhausted our capacity for divine designation. We can utter "God"—and that is all.

And so, if from a point of view of scientific philosophical analysis, God is totally incomprehensible and, therefore, incommunicable, how can God serve any useful purpose to man? Moreover, we are faced with the problem of how man may anticipate the possibility or plausibility of turning to God with the expectation of acquiring a divine revelation for mortal guidance and instruction.

In the hope that there is sufficient motivation to pursue these and related problems to a meaningful conclusion, let us now turn to an examination of some of the inquiries that since time immemorial have ruffled the tranquility of man's placid demeanor, and thereby set aflame his unquenchable curiosity to probe the ultimate domain of human investigation—the habitation of the Supreme Being. For this task, we proceed to a further analysis of "Whence" and "Wherefore."

6

WHAT IS "WHENCE"?

Is a solution to "whence" the absolute goal for man? In embarking upon the ultimate quest, is "whence" the appropriate question?

The distinguished scholar of astronomical science pursues a noble task in striving to achieve a clarification of human

102

purpose and destiny. Sir Bernard Lovell's sincere concern for the welfare of mankind reflects the view of all honorable men of science, who wish to guarantee humanity a safe and sane place under the sun.

However, the arena of investigation permitted by the postulation of "whence" is too limited and too confining. It is a fascinating inquiry, but it stops short of more meaningful targets, and it fails to probe to the significant heart of the matter.

What does "whence" suggest? According to the Oxford dictionary, it derives from Middle English usage, and it signifies: "From what place? From what source, origin, or cause? From, or out of which?" Shakespeare used it in a compound relative sense implying "from where." A cosmological search for "whence" must direct man to the *ultimate* source of all things.

The definitive probe for "whence" demands going beyond "a dense concentration of primeval material" that emerged at the dawn of time. It is, therefore, quite disconcerting to be treated to the spectacular vision of a scientific cosmology, only to suffer the indignity of having the curtains close on the dynamic performance prior to the full disclosure of the intriguing production. It may suit the man of science to consider "a dense concentration of primeval material" as the *ultimate* source of all things: the alert, homespun layman, however, remains dissatisfied, because in his simple naiveté, his commonplace curiosity piques him concerning the essential cause of that self-same "primeval" matter. "Verily," he inquires, "tell us, O great cognoscenti of scientific erudition, what prompted the formation of the 'dense concentration of primeval matter' in the first place? Furthermore, if primeval matter is the *ultimate*, it leaves much to be explained, because it hardly appears to qualify for the role of an *unmoved mover!*"

There is no response. The voice of science is strangely

hushed. One senses the uneasy discomfort of an embarrassed silence on the part of the whole scientific community. It does appear, somehow, that the wrong question was postulated. The curious scientist should not invoke the query "whence." He should, more correctly, limit his inquisitive interest to "how," since science does not step outside the bounds of "how" speculation.

For the most part, science feels duty-bound to assume an official guise of atheism. Individual scientists, who identify God, refer to Him furtively and almost apologetically. Some researchers actually seem to be too embarrassed to mention the Divine Name above a whisper. Many members of the scientific establishment make references to the Supreme Being in a condescending tone. God may be hinted at through heavily veiled esoteric references, but hardly ever is He openly accredited.

One wonders why science, when it reaches the inevitable blank wall at the end of a cosmological safari, does not have the courtesy, if not the courage, to admit that beyond this boundary only God "knows." With all due respect to the magnificent record of scientific achievement, the single notable weakness of science is its display of loftiness in face of the great unknown. Science, all too often, appears oblivious of the fact that its sphere of inquiry is limited to the "how" domain of a Technological Plateau.

When science, therefore, suggests that it is truly content to assume the "dense concentration of primeval matter" as the ultimate source of all things, and that there is nothing beyond it in terms of a "formulator," then science obtusely boxes itself into a severely confining intellectual corner. By regarding such a concretized primal cause as the *ultimate*, science, in effect, makes a declaration to a strange god.

Every person, eventually, establishes his own supreme entity, and this includes the recalcitrant atheist, as well. The ultimate in one's mind is tantamount to an object or

104

concept of deification. For lack of identification of anything or anyone behind the "dense concentration of primeval matter," science remains with an *ultimate* that evolves by default as its peculiar deity. The inexplicable "dense concentration of primeval matter" emerges as the god of the distinguished discipline, and thereby brands science with an absurd pantheism.

The distinguished author of "Whence" effectively delineates the major areas of the cosmological problem, and he appears to suggest the need to search beyond the available subject matter that comes under the scrutiny of scientific analysis. Would it be imprudent to suppose that a leading scientist may also indicate to his peers, who mainly abide by a strong scientific skepticism, that the scope of inquiry beyond matter relates to the realm of God?

At any rate, it is heartening to note the large numbers of humanely disposed scientists who are anxious lest society succumb to an amoral dominion. If civilization is rent asunder by an amoral tidal wave, one could suspect that, in part, it did develop because the populace erroneously determined to pattern its daily life-style after the official amoral posture science necessarily adopts when it is involved in laboratory experimentation.

Responsible scientists who are sincerely apprehensive of human destiny very likely share Professor Lovell's concern for the future of man as the peril of moral bankruptcy spreads. The ethical values of traditional religion are in a state of wholesale decline, and the purported substitutes, such as science and other vagaries of man, have failed to produce a satisfactory measure of existential security or purpose. Humanity languishes under a cloud of existentialist despair; and ironically, this occurs at the crest of civilization's technological triumphs.

The popular pragmatism of the recent past, which has produced our remarkable technological civilization on the one hand, parenthetically has come to grief in affording the human psyche a meaningful sanctuary. While technological man luxuriates in pompous elegance, his alter ego in the existentialist climate endures the misery of profound anguish. Science fails to explain the meaning of life. It does not provide an answer to "whence"; nor can it assume to inquire "wherefore." Indeed, as the learned astronomer avers: from the azimuth of science, it is not possible to "apprehend the ethos of the evening star."

Does this then imply that the attainment of a purposeful, universal ethos forever remains beyond the scope of man? Does this mean that man must eternally be bereft of an unimpeachable moral directive? The sagacious closing statement expressed by the author of "Whence" suggests otherwise. "Human existence," he concludes, "is itself entwined with the primeval state of the universe and the pursuit of understanding is a transcendent value in man's life and purpose."

The acquisition of an uncontested moral law is not an impossible dream. An ethical paradigm radiates from within the bosom of nature. It need only be consigned as the incumbent responsibility of man. Let man pledge himself to nature's ethos, because he is inextricably "entwined . . . with the universe." Man cannot very well survive without professing a sincere commitment to such an ethos, since it represents a "transcendent value" that imparts meaning and purpose to human life.

7

THE "HOW" OF SCIENCE

To attain a modicum of knowledge in the domain of the moral-ethical imperative, man must escape the gravitational constraints of the scientific domain. Science, presently, is restricted to a search for the "how." Scientists, who would soar into the transcendent yonder of more speculative

106

inquiry, must venture into the exotic area of the "wherefore" and beyond.

Momentarily, it appears that the esteemed scholar who asks "whence" is about to formulate the question "wherefore." However, the tradition of the scientific establishment is too strong, and only a feeble allusion is discernible. It is followed immediately by an inevitable retreat to the limitations of the "how."

Precisely, what is the "how"? Through the "how" science undertakes an investigation of the mechanistic or operational aspects of the universe as discharged by the laws of nature. The "how" marks the playground of science. It is a theater of operations that defines and limits the extent of inquiry. Under its discipline, the scientist hopes to elucidate the functional processes in nature. When the chemist discovers the molecular formula, he reaches the climax of his intellectual investigation.

The scientist restricts his field of research to the technical functions of nature. He may be compared to an engineer who comprehends the mechanical aspects of a piece of machinery. For example, an automobile mechanic may know "how" the engine operates, and "how" the vehicle performs; but, he cannot explain "why" any of the components should function as they do—that is, philosophically "why." By the same token, the mechanic is hardly interested in the specific pragmatic application of the vehicle from a moral point of view. After the mechanic completes his repair, he ceases to be concerned with the car's utilitarian disposition. It may be driven by a physician to speed a healing hand to his patients; or it may be used by a scoundrel to commit a crime.

The scientist, similarly, performs his task within the scope of the "how," and abstains from any commitment involving the results of his inquiry. Once he has discovered the "how" of nuclear power, it no longer becomes his concern, as a scientist, as to what may be the manner of its application. The atomic force may become the beneficial mainstay of a community power plant, or it may become the destructive warhead of a nuclear missile.

Incarcerated as he is in a "how" inquiry, perhaps it would be unfair to expect the scientist to express any valid judgment on cosmic cosmology prior to the appearance of matter. The scientist cannot have any explanation for what is beyond the "beyond," since it is outside the range and sphere of his investigations. It appears, in fact, that science, as a discipline, must avoid any adventure beyond the realm of the "how." This may explain why the eminent author of "Whence" hesitates to cross the threshold into the domain of "wherefore," despite the fact that such a step would be the logical consequence of a "whence" inquiry.

It may be quite proper to infer the impossibility of applying the "how" technique of science to ferret out the essence of the "wherefore" from the following:

> The great difficulty is that these evolutionary models for the universe inevitably predict a singular condition of infinite density of infinitesimal dimensions before the beginning of expansion. In this, the theory confounds itself and erodes our confidence in the applicability of the laws of physics to describe the initial condition of the universe.[20]

During the past few centuries, man has made tremendous strides in comprehending new laws in physics and in improving his map of the universe. The entire cosmos is being examined as never before. Yet, there remain secreted within the laws of physics enigmas that may not come to light for centuries. Man is still a long distance away from a comprehensive knowledge of the ultimate potential in matter.

Science encounters impossible problems as it closes in on an evolutionary ultimate. When scientific hypothesis reaches the inevitable blank wall at the end of a cosmological

---

20. Sir Bernard Lovell, Supra, p. 34.

journey, its analytic methods become inadequate. Science can conceive a neutron star with matter of such extreme density that a mere speck would weigh a million tons. Some astronomers consider that neutron stars may explain the nature of the mysterious pulsars, which emit radio signals of extraordinary quality. However, the true nature of these strange stellar spectacles remains in the domain of the unknown.

Consider, now, the utter perplexity confronting the astronomer when he postulates an "infinite density of infinitesimal dimension," an imperceptible concept many times more complex than the manifest neutron stars. How can he approach such a staggering intellectual prospect? Pathetically, the astronomer comes to an intellectual impasse, and he must throw up his hands in hopeless despair, because the "how" process of science disintegrates as it nears the infinite.

Not only is the "how" an ineffective implement as it approaches the threshold of infinity, but it becomes totally useless past that point. The method of the "how" cannot operate beyond the universe of matter. The "how" collapses when it is exposed to the stress of what may turn into an abstract investigation, and such a destination appears imminent when the directional arrow of "whence" is posted. The "how" manner of research erodes as it approaches the *ultimate*.

In order to proceed, the intellectual investigator must transfer to a new line of inquiry, and the next tool on the inquisitive hierarchy is "wherefore." The "wherefore" permeates the intellectual area outside the domain of science. As the man of science leaves the "how" to pursue the "wherefore," he must also shed his laboratory frock and identification card, so that he is no longer the scientist in search of concealed physical knowledge, but another personage.

INTRODUCING "WHEREFORE"

What is meant by "wherefore," and who, if not the scientist, may indulge in its exploration? For its literal definition, we turn to the Oxford dictionary, where the term is listed as indicating: "For what? For what purpose or end? For what cause? On what account? Why?" Simply stated, "wherefore" asks "why."

There are some who proclaim that man does not have the right to indulge in a "why" inquiry, because it relates to secrets beyond the human ken that had best be left undisturbed. It would seem, however, that since man has the ability to formulate abstract, metaphysical questions, it becomes his obligation to pursue their solution to the maximum capacity of his intellectual ability. If this premise is acceptable, then we may turn to the next item on the agenda, namely, what manner of individual may become involved in a "why" exploration?

Of course, there are those who may ask: What does it matter if one examines the question or not? Well, if one presumes to be a cosmologist, as in the case of advanced scientific thinkers, then the route of cosmological investigation must be probed to its logical climax. The proper researcher must not cease in the midst of his investigation. Neither the eminent astronomer nor any other astute scientist would dare to stop midway in the course of an important experiment, and record the results discernible at that point as being the final conclusions in the matter. So is it with cosmology. One cannot limit a cosmological analysis to the "how" plateau alone. A consideration of all questions beyond the level of the "how" is an essential responsibility of the cosmologist. Such a program of interrogation must be followed even if it calls for an eventual divestment of the utilitarian garments entailed in "how" procedure according to strict scientific method.

THE EIGHT LEVELS OF HUMAN EXISTENCE

There are eight basic personality denominations that an individual may assume in life. This division dates back to antiquity. It developed early in human history, at about the time when man took his first step across the threshold that marked the birth of civilization, somewhat less than six thousand years ago. Man may identify with any one or more of these categories, and his culture may likewise reflect an association of similar characteristics. At times, both man and his society may fall under the sway of one trait above all of the others, so that it completely dominates the communal life-style.

The eight categories are grouped under two basic divisions, prosaically designated as the *Biological Plain* and the *Intellectual Plateau*. Each domain exhibits four sub-divisions. Within the Biological Plain, there roam the likes of Brute Biological Man. Domesticated Biological Man, Civilized Biological Man, and Majestic Biological Man. The latter is segregated from the others, and secures his domicile on the luminous side of the Intellectual Plateau. Majestic Biological Man finally arrives at his idyllic homestead only after he has painstakingly concluded an arduous trek across the Intellectual Plateau.

In its hierarchal design, the Intellectual Plateau marks the ascendency of man through the gradations of Technological Man, Philosophical Man, Theological Man, and Prophetic Man. At each level, man develops the unique characteristics identified with that particular personality.

As man hovers over the entire panorama of the Biological Plain and the Intellectual Plateau, he is ordained with an unusual degree of freedom to select the existentialist area with which he desires to be associated. Freedom, then, alludes to man's inclination and ability to choose an identity from among the eight varieties of potential human existence. A mobile personality may develop as the individual fluctuates from one domain to another domain and from one level to another level, in the course of a lifetime, or in the course of a single day. These peregrinations, which vary in accordance with personal predilections, formulate the existential ego noesis.

On each level, man endeavors to comprehend the nature and manner of the world around him. Generally speaking, Biological Man is cloaked in the folds of nature, and his vision is limited to the luxuriant panorama of his immediate environment. His attention is riveted upon the mysteries of nature at his doorstep. He seeks simply to know the habits and mannerisms of the fascinating array of specimens with which he comes in contact. His mind is attuned to the "whatness" of nature, and his question entails the "what" of the subject. "What is this?" he asks.

But now, let us turn to a more precise enucleation of man as he emerges from the kaleidoscope of mutable identities. The gradations bring into focus an interesting display of personalities.

In the nadir of the Biological Valley, *Brute Biological Man* wallows among the self-centered, unfeeling, illiterate slime pits. His primitive ignorance exudes an elementary, superstitious, amoral existence. A tribe of such creatures may survive by virtue of instinctive mutual tolerance. Non-tribal interlopers may fall prey to the brute's nocuous malevolence, and they may be dispatched without so much as cause or provocation.

Farther along the coarse Biological Plain, one finds *Domesticated Biological Man*, who pretends to be the manifestation of an urbane civilization. Despite his ostentatious display of cultivation, this specimen still adheres to the amoral convictions of his nether brother, Brute Biological Man. The posture of comity and obeisance to legal standards exhibited by this domesticated version of human-

112

ity is merely a costume. By exhibiting this facade, he treads a narrow path on the decorous side of a rigorously patrolled society, where infraction of the law is severely punished. Given the opportunity, however, he would have no qualms about violating a restrictive civil code. In fact, he could relish an indulgence in a torrent of perverse behavior that may surpass the inhumanity of the brute.

On a slightly verdant Biological Plain near the approach to the Intellectual Plateau resides *Civilized Biological Man*. He exhibits the social grace, the superficial amenities, and the amiable suavity of an artificial culture. Although he subscribes to the collective patterns of a respectable society, he remains an irritant to its stability. It is not so much what he exhibits by way of formal conduct that may ruffle the placid scene, as what he fails to express. He may display honesty, but he lacks kindness; he may expound fairness, but he ignores fraternal tolerance and sympathy. Since he is dominated by the egocentricity of a self-serving biological nature, he constitutes a potential hazard to the successful perpetuity of a civilized society.

The fourth dimension of the Biological Plain lies beyond the Intellectual Plateau. Lush green meadows and sylvan gardens of arboreal splendor abound in this domain. A veritable botanical paradise is the abode of *Majestic Biological Man*. Why should he not repair to such a delightful domicile? After all, he has diligently climbed all of the precarious levels of the Intellectual Plateau to gain the enlightened legacy radiating from the summit. He embraced tenets of profound sagacity before he earned the title of Majestic Biological Man. He is honest, kind, and innately moral. He is dedicated to the highest ideals of civilization, and he is concerned with its welfare. He is considerate of others; regards them with a true sense of empathy; and he is mindful of all his obligations. Above all, he is God-fearing. If anyone deserves to experience the essential quality of happiness in life, Majestic Biological Man ranks in the forefront of prospective candidates.

Turning to the Intellectual Plateau, it rises handsomely before us as the more intriguing division of human existence. It consists of a series of subplateaus, each ascending as an acclivity from a lower slope. Here is the region where man endeavors to solve the problems of life; where he seeks to unravel the origin of things; where he searches for a more sophisticated explanation of cosmological riddles.

On the Biological Plain, man observes; on the Intellectual Plateau, man considers. On each grade of the Intellectual Plateau, man puts forward challenging questions, and in accordance with his mental perspicacity, he establishes definitive conclusions. The panorama of man runs its full course through this commonplace cycle: man observes; man considers; man concludes.

The queries that man formulates during his upward trek on the Intellectual Plateau are motivated by a desire to achieve a rational explanation of the seemingly endless enigmas emerging from nature. Upon being confronted with an array of unsolved perplexities, curious man feels compelled to seek clarification in the realm of the abstract. He is obliged to ask four basic questions: How?—Whence? —Wherefore?—Who?

Technological Man pursues his inquiry on the application level. He perceives a regularity, or a design in a specific operation of nature, and he commences to recognize a pattern that makes possible its future predictability. The reliability of matter's behavior enables man to develop a discipline of science.

Technological Man is committed to a search for knowledge in the area of cosmography, so that he may produce a description of the structure of the world. Numerous disciplines rise to the task, and the research investigator becomes a major figure on the Technological Plateau.

Contrasted to the "what," or simple operation of nature, which is the chief interest of Biological Man—exemplified by the hunter, the woodsman, the farmer—on the higher level, Technological Man endeavors to comprehend "how" nature performs. He wants to know "how" matter is constituted and "how" it is formulated. The "how" of water is satisfied when it is determined that it is composed of a specific combination of hydrogen and oxygen.

Both Biological Man and Technological Man share in common a proclivity for amoral projection. The ego on the lower Biological Plain rejects a responsible relationship with any other ego. Biological Man is totally self-centered. He is quite remote from God, so that even a conceptual apprehension of the Supreme Being is alien to him. His base, menial outlook leads him to identify with fiendish, idolatrous cults.

Technological Man's amoral position stems from an unfeeling attitude toward otherhood. This posture is commonly exhibited in the laboratory of experimentation. The dedicated scientist may regard the objective of his research as of uppermost importance, and therefore, the subject undergoing the research may be treated without the slightest tinge of emotion.

Technological Man is quite self-centered. That which is outside the periphery of his ego is considered as an "it." Both man and matter may be treated as an "it"; and if they are categorized merely as an "it," they become readily expendable in the cause of technological progress. In searching for the "how," the *ends* may often justify the *means*. Consequently, Technological Man may adopt an ignoble pragmatism as his criterion for deriving values in life. His objectives are then pursued with a ruthless efficiency, and he applies to them a pragmatic materialistic philosophy grounded upon the crude doctrine: If it works, it is true!

The amoral costume of Technological Man obliges him

to avoid the "why," since it opens the door to a moral challenge. Technological Man feels that he cannot survive the penetrating glare of moral exposure.

Philosophical Man gains an awareness of abstract concepts. God, as an abstract idea, emerges on the Philosophical Plateau. Neither Biological Man, nor Technological Man achieve a proper comprehension of God, because to do so, it is necessary to appreciate the full scope of the formulary universal implanted within the body of nature. Philosophical Man recognizes the universal "Idea" as the product of a Supreme Intellect. Once the "Idea" is fully cognized, it may become a "point of contact" between the mortal intellect, which perceives the "Idea," and the Supreme Intellect, which formulates the "Idea." If such a juxtaposition were exploited to its fullest advantage, Philosophical Man could establish an abstract confrontation with God. He avoids such a communion, however, because a communicable God would thrust too many moral responsiblities upon his shoulders. Philosophical Man wants to keep God out of reach, if not out of sight.

Philosophical Man surpasses Technological Man in searching for meaningful answers to life's riddles. Philosophical Man sees the universe as an ordered whole, and so he attempts to invest nature with significance. He seeks to identify the reasons and explanations for nature's unique behavior. Philosophical Man wants to know the "why" and "wherefore" of everything.

In Philosophical Man's cosmological atlas, God emerges as the transcendent source of the universe. He appears as the *ultimate*. His intimate being, however, still eludes Philosophical Man. God is regarded as a third-person identity—as a remote "He." The allusion to God as a distant abstract concept precludes the possibility of establishing a personalistic relationship. Philosophical Man has no interest in an individual affiliation with God, and, there-

116

fore, he pays little heed to His identity as a Divine Personality. As a result, Philosophical Man considers it futile to attempt any manner of communication with God. You cannot talk to a "He." If God is only an abstract concept, and not a Divine Personality, there can be no dialogue between Him and man. As a "He," God cannot be petitioned. Philosophical Man finds prayer fatuous, and so he does not pray.

Limiting God to a remote abstract concept produces a more serious consequence. If God is viewed as a distant "He," then the moral norm loses its binding quality for Philosophical Man. When God appears only as a "He," any divine moral imperative, which may be discerned by man, likewise exhibits nothing more than a third-person relationship. The divine moral law becomes a distant "it," and as such, is divested of the binding character that is necessary for encouraging a personal commitment.

Philosophical Man resides on a deceitful plateau. Abstract concepts, such as *God*, the *ethical norm*, and the *moral imperative*, may be frankly discussed and fully analyzed, yet they may be completely ignored as personal obligations. Since God remains beyond the scope of human communications as a "He," and since moral postulates are examined as an objective "it," they are discretionary matters for Philosophical Man to ponder upon, but not necessarily to accept as an incumbent imperative.

On the Theological Plateau, man discerns moral reflections within the broad stream of natural law. These moral vibrations inspire a mood of appreciation for the sagacious, purposeful structure that nature exhibits. The apparition of an orderly system embracing a cohesive scheme of causal development and processes impels *Theological Man* to move beyond the realm of "why" speculation. He is confronted with metaphysical challenges that demand the formulation of the question "who."

117

Theological Man rejects any suggestion that the meticulously structured universe evolved haphazardly from blind, indeterminate matter. For him, such a notion is contrary to all the basic elements of human logic. Only a *persona* of omnipotent ability could be credited with drafting the magnificent blueprint for a well-regulated, complex cosmos. Theological Man concludes that a rational cosmology can only be developed by searching for a "who" behind the cosmic facade.

In hoping to identify the almighty "who," Theological Man embarks upon an exploration for a "thou." The God of creation, who brought the universe into being, and who invested it with a glorious *modus operandi*, becomes more accessible if He is endowed with a second-person frame of reference. As a "Thou," God is much closer to man than as a "He." This is an important distinction. The God to whom Philosophical Man refers on the lower plateau is narrowly restricted to a remote third-person category. Philosophical Man considers that a mortal ego can only speak *of* God, since God is identified as a distant "He." Theological Man seeks a more intimate relationship. Therefore, the mortal "I," as viewed by Theological Man, may speak directly *to* God, because God is contemplated as an accessible "Thou." It is possible to address a "Thou," but not a "He."

An "I-Thou" relationship between man and God becomes possible on the Theological Plateau. By investing the Supreme Being with a Divine Personality, which is what a "Thou" implies, God assumes the more intimate apparition of a Divine Being. Theological Man concludes that communion with the Divine Being is not only possible, but moreso, highly desirable. Man need only discover the proper wavelength for engaging in such communication.

Theological Man addresses the Divine Personality as the blessed "Thou" of prayer, hoping that the one-way effusion

118

of adoration may elicit some sort of response. The frenetic search for a divine reply entices Theological Man, at times, to interpret some unanticipated act or omen in nature as an expression of divine communication.

The more sophisticated Theological Man realizes that the expectation of a divine replication at the behest of man is futile. Instead, sophisticated Theological Man peers more deeply into the recesses of nature to locate additional clues, because these may reveal more of the glorious aspects of the Divine Personality. Such a search may enable man to discover new, meaningful facets of the divine universal scheme. Despite the handicap of a finite meager intellect, man may boldly attempt to discern a divine exhibition of will within the confines of nature, which is tantamount to the reverberation of a divine rejoinder.

The voice of God resounds through nature. Moral vibrations emanating from nature reflect cohesive aspects of the divine plan. These may be explored in greater detail on the Prophetic Plateau, where they may emerge as the basis for formulating an addendum to what has already been revealed as a divinely inspired ethical code.

Although sophisticated Theological Man precludes the possibility of a divine answer, he, nevertheless, is inspired to indulge in prayer, because he is psychologically stimulated to enact his role in an "I-Thou" dialogue. Prayer is an ennobling experience for Theological Man. It exhibits a prodigious confidence in the ability of the human ego to reach beyond the universe and contact the Divine Personality. In perceiving the Divine Personality as a "Thou," man is filled with an odylic passion to express his innermost thoughts unto the Divine Presence, even as one humbly beseeches and petitions a mighty sovereign. The sovereign need not deign to reply.

At the summit, *Prophetic Man* likewise pursues the enigma of the "who," but he is determined to discern the identity of God in a more personalistic appearance than the one revealed to Theological Man. Prophetic Man perceives God not as a "Thou," but as an "I." God, in order to enable a deeper encounter with mortal man, appears to Prophetic Man in a first-person guise. The Divine "Thou" becomes a Divine "I."

Upon the emergence of God as an I/Divine, it may be encouraging for man to seek a confrontation by raising his own I/mortal being to the status of Prophetic Man. The Prophetic Summit permits man to contemplate God in an enlightened cognitive attitude, with the hope that he may acquire the intellectual skill to formulate a two-way dialogue. The chasm between the infinite and the finite may be bridged by establishing a point of reference, which would enable the Divine Ego to engage in an apperceptive revelation with a mortal ego. By raising the ego awareness to an "I" level on both sides, a communion between man and God becomes possible. God is no longer a silent "Thou." The Divine "I" addresses man. It is a profound moment in the human experience. Prophetic Man finally discerns the supreme declaration of the I/Divine: "I am the Lord, thy God!"

How does man arrive at the noble stature of the Prophetic Summit? Man ultimately conceives the mortal ego as the unique cradle of a distinctive, conscious personality, which provides him with an awareness of the "other." On the level of human relationships, the "other" is cognized as fulfilling an ego-ordained existence. Meaningful communication can be established when two I/egos appear in contraposition.

Man notes that only an I/ego speaks; a "thou" remains mute. The Creator is considered to be no less deficient than the most ingenious intellectual product on planet Earth—man. If the human intellect perceives its own ego-consciousness, shall such an ability for cognition be denied God? The unfoldment of the mortal I/ego certifies the manifestation of an I/Divine. In this light, the human I/ego personifies a reflection of the Divine Ego, so that it may be said that man is made in the image of God. Both God and man possess a singular "I." By unveiling the I/ego of the Supreme Being, God becomes accessible as a potential respondent in a two-way dialogue.

Of course, the Divine Ego is extremely obscure, and the means for comprehending its nature are beyond human capacity. There is, however, one avenue of revelation open to man. Intellectual Man may detect and apprehend clues reflecting some aspects of the Divine Personality by studiously discerning them within the Creator's masterful production—nature. An analysis of a marvelous mural may disclose bits of the personality of the unknown artist. By assessing the full scope of nature and drawing therefrom reflections of moral qualities, man may hope to derive meaningful concepts expressing the will of God. For example, nature is dedicated to the principle of producing and preserving life. The preservation of life and existence becomes a moral norm, particularly when the subject exhibits a conscious cognition of the ego within, as does man. A simple corollary suggests that the destruction of these unique specimens of ego-conscious life, who have the ability to perceive God, is a negation of nature's plan, or God's will. A moral precept emerges: "Thou shalt not kill!"

An ego may be discerned by comprehending its will. It is a difficult challenge. The querist strives to gain an awareness of the inaccessible artist's personality by studying nought but his masterpiece, because that is all that is at hand. More demanding is the endeavor to elicit the moral predilections of the Creator through an analytical examination of his *chef d'oeuvre*—the universe. Man must rely upon an appropriate interpretation of the concealed personality radiating from within the bosom of nature, in order to fathom

121

selected ethical precepts reflected by the Divine Ego. The emerging moral dictums are no less significant an experiential relationship than the data discerned by the scientist, except that Prophetic Man conceives his revelations as divinely ordained declarations. For this reason, an encounter between the I/mortal and the I/Divine comes to pass only on the Prophetic Summit.

After he has accumulated his precious information, Prophetic Man feels compelled to share it with others because it represents a divinely revealed message. He may be inspired to proclaim it as the enunciation of a Divine Law; or as a promise of blessedness; or as an exhortation against waywardness. In whatever manner it is formulated, Prophetic Man experiences the deep anguish of responsibility to convey to his fellowman the revelation he elicited through an I/mortal-I/Divine confrontation on the Prophetic Summit.

Prophetic Man addresses himself to the intractable masses huddled on the lower existential echelons, and he promulgates the divine will in accordance with his cognition of the Divine "I." He hopes, thereby, to inspire the multitude on the plebeian levels to pursue the paths leading to the Theological Plateau, because it is a noble stage of life that offers the most desirable prospect for authentic existence. Authentic existence is within the grasp of every individual. One need only direct his intellectual powers toward the Theological realm, where an I/mortal-Thou/Divine encounter may take place.

Of course, the road to authentic existence upon the Theological domain is not strewn with roses. Man must undertake a challenging journey across the entire Intellectual region prior to his ascension to the Theological heights. But, it is well worth the effort, for such a lofty Theological perch unfolds unto man's vision the verdant brilliance of the Majestic Biological Plain. Stepping from the Theological

122

Plateau along a highway to the Majestic Biological experience, man can learn to appreciate anew the promise of a meaningful existence in a mortal cosmic universe under God.

Society, itself, may be identified by the domain it enshrines. The mores of human culture are the criteria for computing a people's status. There are provinces where life does not reach beyond the lower Biological Plains, and there are states that accede to Technological supremacy as the favored way of life. It should surprise no one if these societies soon adopt a brutal, amoral polity as their key to survival. In some instances, a culture may be attuned to the Philosophical level. Its population could very well reflect a serious attitude toward all things. On rare occasions, a civilization may gain inspiration from the Theological highlands, and its people will be devoted to the maintenance of a noble, moral fraternity. Such a community cannot be blamed for exhibiting a determined reticence toward any involvement or fraternization with societies pursuing lower categories of existence. Its aloofness is justified by the fact that any intermingling carries with it the danger of moral pollution. The purified atmosphere of the "who" on the Theological and Prophetic heights may readily be despoiled by the infiltration of rank abominations from below.

10

THE PARADOX OF MODERN SCIENCE

As Professor Lovell indicates, cosmologists may arrive at a point of inquiry that lies beyond scientific comprehension. It is a condition that eludes the applicability of known physical laws, and renders futile the attempt to externalize defiant problems.

Contemporary cosmological speculations are rife with intellectual quandaries. Many of these enigmas stubbornly elude the analytic perspicacity of the scientist, because quite simply, twentieth-century man lacks the necessary fundamental knowledge to comprehend and deal with the full scope of such investigations. Despite man's fantastic imaginative capabilities, he finds himself severely limited in pursuing theoretical assumptions. At every turn outside the sphere of recognizable phenomena, there arise such a vast array of formidable barriers that it becomes practically impossible for the human mind to hurdle these intellectual challenges. Yet, the indomitable human mind refuses to be stifled, and it forges into areas of forbidden speculation with the courage of an emboldened gladiator. Of course, the logical human mind must operate within the bounds of logically recognized specifications. As a result, when the scientist assumes the prerogative to assert a theoretical proposal, he must proceed in accordance with rationally established principles heretofore certified. For the cosmologist, this poses a problem, since it severely restricts his imaginative entry into the realm of the unknown. Nevertheless, these areas cast a spell of intriguing allure, and promote a variety of speculative concepts.

Cosmology in the twentieth century radiates mostly between two schools of thought: the open-universe theorists; and the closed-universe theorists. These views may derive from an acceptance of the big bang theory as a plausible explanation for the beginning of our universe. The big bang is presumed to have taken place approximately ten thousand million years ago, and according to some cosmologists as early as twenty thousand million years ago. Thanks to the brilliant work of Edwin P. Hubble, it became possible to suggest the velocity of galactic recession, thus enabling a calculation for dating the big bang. The figure that Hubble computed for the age of the universe was two thousand million years; however, geological indications pointed to a much older planet earth, sun, and universe. Walter Baade's application of the two hundred-inch telescope at Palomar helped to re-establish a figure of at least five thousand million years for the age of the universe. Since then, the value of the Hubble constant has been several times amended, and the consequence has been to assume an older and older age for the universe.

Whatever the date for the onset of the big bang, the consensus appears to be that in the beginning an extremely dense torrid mass of primordial matter exploded, sending gases flying out into the emptiness of space. Some of this gas cooled and contracted, forming the galaxies and stars, but their outward expansion nevertheless continued. The important question currently facing cosmologists is the ultimate consequence of such expansion. Will the universe eventually reduce the speed of its outward flight to a point of total cessation, after which, it will reverse the entire process and fall in upon itself to reappear as its original dense primordial ball of matter? To produce such a closed universe, it would be necessary to ascertain a slowdown or deceleration of galactic expansion of sufficient proportions so as to enable the cosmic mass to experience the necessary gravitational braking force to bring on the reversal process. On the other hand, measurements have been produced to suggest an open universe in which the galactic expansion would continue eternally in infinite space. One senses an important philosophic turn of mind as a result of an identification with either a closed or open universe.

By using the red shift as a determining factor for assessing the rate of galactic recession, Allan Sandage of the Palomar Observatory estimated the life span of the universe to be eighty thousand million years. This projection was made in 1960 taking into account the whole range of cosmic existence from the moment of the big bang, through galactic expansion, and back to the primordial fireball condition. Apparently, Sandage felt that his conclusion was somewhat premature, because in 1974, he adopted a new figure for the Hubble constant that necessitated a change in the considered age of the universe by some sixteen thousand million years. He also questioned the slowdown of galactic expansion to which he was committed in his earlier speculations. In contrast, his new calculations indicated that the universe would expand continuously, and he thereby adopted the cosmological view of an open universe. In this assumption, both he and James E. Gunn of Caltech, who also concluded an open universe concept, were confronted with a substantial paradox. Their cosmology entertained the notion of a finite universe committed to a never-ending expansion of infinite proportions.

Of course, the closed universe proponents stoutly defended their point of view by maintaining that astronomical information is not always accurate, and in fact, it may be subject to a variety of interpretations. Despite any apparent difficulties, the closed universe emerges both logically and philosophically as the more appealing cosmic design. This decision is based upon a further assessment of Sandage's current cosmological position.

Although Sandage has changed his view on the deceleration rate and the availability of cosmic gravitational matter that would make an ultimate galactic recession possible, does he not yet acknowledge a *slight* deceleration along with the luminosity-red shift relationship that is identified with a uniform cosmic expansion? If such is the case, does not even the faintest hint of deceleration at this moment in cosmic history pose a problem for the open-universe advocate? On a universal scale, it appears logical to assume that once a deceleration process has been set in motion, such a process would be likely to increase in time, and ultimately it should arrive at a checkpoint for any further galactic expansion. Does not a common-sense judgment suggest that the admission of a minor deceleration, slight though it may be, conflicts with the notion of a negatively curved hyperbolic space accommodating a violent, eternally expanding universe? At best, the open-universe theorist may interpret a mild deceleration as accommodating a flat geometry wherein galaxies may come to rest after they have separated to an infinite distance from each other. But, how can such a cessation of galactic expansion ever come to pass? If, in a proposed flat geometry of space, the galaxies will presumably finally "come to rest" when they have flown infinitely apart, how, indeed, can they ever "come to rest," since as long as they remain in a corporeal state of matter, they will never complete their space-time journey of infinite separation?

It is enormously difficult to comprehend a theoretical graph of the Hubble relation for distant objects in its demonstration of a flat curve, implying an ultimate "coming to rest" of galaxies when they have gained infinite separation. It seems that the "coming to rest" will never be achieved, because matter in the form of material bodies can never reach a state of being "infinitely apart." Consequently, since the theoretical graph of a universe reflecting a flat geometry depicting stationary galaxies at a point of infinite separation can never be realized, any suggestion of such a flat geometry must eventually emerge as nothing more than a masquerade for what is in reality a never-ending energetically expanding universe corresponding to a negatively curved hyperbolic space, such as mentioned above. However, in the above-cited instance, it was indicated that upon the assumption of any deceleration, no matter how slight, the ultimate consequence must lead to a retreat from the concept of an infinite galactic expansion. Accordingly, since the flat geometrical space concept appears to be but a redundancy of a negatively curved hyperbolic space, there appears to be no other alternative for accommodating the fact of galactic deceleration other than assuming the third remaining geometry of space, namely a positively curved spherical space that points to an oscillating or closed universe. If Sandage admits to a deceleration, slight though it may be, the structural pattern that seems most logical for a deceleration factum is not an open universe, but rather the closed model with an oscillating potential.

There is another area where the paradoxical aspect of nature puzzles science. The problem occurs in the speculations attending the final stages of stellar collapse. In pursuing this enigma, astronomy has propounded the fascinating concept of what is now called a black hole. When a star has exhausted its energy, it commences a routine of dynamic change eventually waning to assume the identity of a white dwarf. At that point there is a question as to whether Einsteinian general relativity could tolerate further condensation. Sir Arthur Eddington concluded that a star may reach a point of contraction wherein its gravitational field became so strong that its own light rays would be trapped so that it could never again be seen. After a star collapsed to a tightly packed neutron stage, any further contraction would produce the scientific anomaly of a black hole.

The black hole concept defies presently known laws of physics. It is a phenomenon entailing an infinitely gross mass and gravitational force compacted within an infinitely small space. It would no longer resemble any aspect of matter as we commonly know it. All of the atomic electrons and protons would have been crushed into neutrons, and the further order of collapse would proceed to condense the star into even smaller entities. The notion of a space-time continuum would be utterly dissolved; space would become a futile term, and time would elude comprehensible measurement. If the concept of relativity, which conceives gravity as a curving of space-time, were applied, the space-time curvature would be totally enmeshed within the confines of the unseen star. Any matter encountering a black hole

128

would experience immediate annihilation. Indeed, here is a paradox of physics that appears to exceed the human limits of visual and mental comprehension.

Along with the intriguing enigmas produced within the confines of nature, there arises a quandary relating to the cosmological position assumed by some of the notable contemporary scientists themselves. Generally speaking, the scientist of today is a paradoxical figure. On the one hand, he recognizes an incontrovertible design in nature, and he may endeavor to assuage his dilemma arising from the appearance of a persistently predictable regularity in the cosmos by convoking conferences to explore the consequences implied in admitting to such a design in the universe. On the other hand, the very same scientist firmly denies the existence of any "designer"—divine or otherwise. Such a contradictory posture appears to be rather illogical. Yet, scientists persist in sponsoring a viewpoint that blows hot and cold in the same breath.

Of course, the scientist professes to have reasons for denying the Supreme Creator. Darwinian "natural selection" is a handy weapon with which the scientist may counter logical queries. Life apparently exists in abundance all over the universe, but the scientist considers life as being indubitably pledged to a principle of design called "natural selection." The scientist further contends that "natural selection" does not ascribe to a technological process, but rather to a unique method, which by logical standards defies human credulity. "Natural selection" develops without relating to any prior specifications whatsoever.

Why must the scientist deny a technological process in nature? The answer is quite elementary. By cancelling a technological process, the scientist hopes to avoid the challenging question "who." Despite the fact that everything we see suggests a technological design, and everything we produce subscribes to a technological methodology, the determined scientist would have us believe that Mother Nature baked all of her two million and more ecologically balanced and self-propelling species of complex confections, without once ever having glanced into a cookbook.

How was this remarkable feat accomplished? The scientist offers a simple explanation. "Natural selection" operates in accordance with three components: (1) a ceaseless outpouring of variations; (2) a mechanism of inheritance; (3) a selective factor, or a competitive principle, which Darwin called "the struggle for existence."

There is one major problem with the theory. It considers that, by endeavoring to explain the "how" exclusively, it successfully demolishes the area of the "why" and the "who." Orthodox Darwinians may suppose that they can pull the wool over the eyes of Mother Nature, but they certainly do not satisfy the intellectual curiosity of man. The loopholes that science knits into its theoretical cosmological fabric are exceedingly large and plentiful.

First of all, matter commits itself to an immutable existential paradigm that took effect at the primordial split second of its earliest appearance. If it would have failed to adhere to an assigned ultraprecise identity, our universe could never have come into being. Presently, the essence of matter comprising the electron and proton, conceives the proton as containing almost two thousand times the mass of the electron. The nature of these two elementary particles poses a paradoxical problem for science. Although they are proportionately extremely different, nevertheless, they carry the same numerical charge, if but opposite—the proton is plus; the electron is minus.

Was this an accident? Did this occur by chance? There is more to the story. Every good physicist knows that if the proton and the electron differed numerically in their charge, everything would have been charged. However, there would have been one severe drawback—a universe, to be so chargeable, would never have evolved in the first place.

In early 1973, it was determined that the equality of the proton and electron charges differ in so small an infinitesimal quantity as to be negligible. It is due to this very sensitive and delicate balance of the nuclear forces that a universe can develop through a process of expansion, and through a further process of condensation. If the single proton and the single electron of the hydrogen atom varied by the slightest iota from the formulated equality of charge, there would be no gravitation; there would be no stars; there would be no galaxies; and there would be no questions for man to ponder, because there would be no man.

In his admirable essay, Sir Bernard Lovell establishes this premise as an indispensable determinant for the emergence of our universe. The statutory facts as cited in the following excerpt are highly significant:

> Indeed, I am inclined to accept contemporary scientific evidence as indicative of a far greater degree of man's total involvement with the universe. The life which we know depends on a sensitive molecular balance; the properties of the atoms of the familiar elements are determined by a delicate balance of electrical and nuclear forces. These and the large scale uniformity and isotropy of the universe were probably determined by events that occurred in the first second of time . . . the existence even of stars and galaxies depends in a delicate manner on the force of attraction between two protons. In the earliest moments of the expansion of the universe . . . if the proton-proton interaction were only a few per cent stronger then all the hydrogen in the primeval condensate would have turned into helium in the early stages of expansion. No galaxies, no stars, no life would have emerged. It would have been a universe forever unknowable by living creatures. The existence of a remarkable and intimate relationship between man, the fundamental constants of nature and the initial moments of space and time seems to be an inescapable condition of our presence. . . .[21]

Noteworthy, is the conclusive acknowledgment that man and nature are inextricably bound by the common ancestry of their genesis. Do they not, then, share alike the destiny

---

21. Supra, p. 36–37.

of an ultimate eschatology? Furthermore, if nature bears the imprint of a divine design, then man, too, being totally involved with the universe, is divinely ordained; and the impelling desire to seek his Maker is not a futile gesture. Man may, therefore, arduously search for the voice of God, which echoes throughout the expanse of nature.

If, however, it is assumed that nature has no divine sponsor, then the scientific genesis story is fraught with logical inconsistencies. There remain unexplained questions that lead to a cosmological impasse. For example, were the precise properties that were adopted by the proton and the electron the consequence of an "accident" or "chance"? Did the hydrogen atom become attached to the Darwinian clothesline of "natural selection" all by its lonesome self? It surely appears that these invisible particles of matter would have to be pretty brainy creatures to have figured out such complex, sophisticated permutations, that they could know what manner of precise physical identity they must assume in order to produce an expanding and condensing universe. In addition, let us not overlook the technological process. If the definitive behavior patterns assumed by the early identifiable particles, as we now comprehend them, do not exhibit a "technological" design, then perhaps our definition fount has gone awry.

What Darwin identified as the process of "natural selection" pertains to the whole domain of the Biological Plain. Science acknowledges that our universe is designed to breed life. It further asserts that to think in terms of a "designer" would be a technological concept, not an organic or biological idea.

One wonders how science could blithely ignore the fact that nature, through "natural selection," does not create its own variations, and that these are predetermined in accordance with the original properties with which primordial matter was endowed "in the first second of time." To

132

credit "natural selection" with the formulary creation of a complex universe is tantamount to subscribing to an unpalatable pantheism of absurd dimensions.

If science deems it proper to assume a near-sighted position by proclaiming nature as the progenitor of "natural selection," this would, at best, make nature an editor-in-chief, but it would hardly qualify nature for the position of Master-Planner or Creator. Matter, both inorganic and organic, is eternally pledged to a formulary design indicative of universal schemata emanating from an intellectual milieu. If the scientist alleges that amended conditions produce strange, unknown variations, it seems only logical to regard the resultant mutants also as preordained developments of the original formulary blueprint, when its reproductive design process is deviated. The emergent conclusion points to the fact that the whole cosmic process is not a chance evolvement through blind "natural selection"; rather it is more credible to view the cosmic panorama as the foreordained progression of a "discriminating creative process" directed by a majestic Prime Mover.

Despite the logical appeal of this position, one wonders what it is that motivates the scientist to try to avoid a confrontation with the "who" so desperately, that he must resort to a spurious dichotomy that he hopes will distinguish human technology from natural technology? Is man so vain that he can brook no competition in technological contests? Is he zealous, lest his supremacy as a producer be by-passed by the copious fecundity of nature? Man can fashion nothing unless he utilizes the materials and resources of nature. Man can only exploit the potential in nature by forging, constructing, shaping, or chemically refining matter a little differently, so that it is constituted in a design and form more immediately useful to him. In the process, however, man has produced nothing new. Every human fabrication and ingenious invention is merely the artifact that man produces as a result of his alert analysis of the *preinvested* properties of matter and its remarkably flexible versatility. Without nature, man can bring forth nothing. The term *man-made* is a gross misnomer.

Moreover, man is immured by nature's laws. Man does not make matter do his bidding; he merely guides its innate behavioral traits so that it may enact a new role in a revised script. The play, or the total production, however, is the same, and it all comes under the grand technological description of "Nature's Follies," with man quite often presuming to be its star.

Let not the scientist assume the imperious stance that *discovery* is the equivalent of *creativeness*; nor should he indulge in arrogant sophistry by suggesting that in all the universe only the brain of man rears up as the supreme intellect. Such haughty self-adoration leads to self-deification, and with the destructive potential man now possesses, he cannot afford to gamble away his chance for survival by betting on the wrong Supreme Power.

Science need only evaluate its own conclusions, and it would unerringly be able to postulate the *true* Supreme Being; but this would require a movement upward to the higher intellectual plateaus. Yet, even from its present vantage point, it is possible to discern identity strains pointing to the *ultimate*. Consider: the very first appearance of a charge of energy was simultaneously enacted and determined in accordance with set and established physical law. In the beginning, there was law! This, no scientist in the world can deny. The corollary is likewise beyond challenge. We cannot possibly conceive of a law without logically assuming a Lawmaker. But why use the term *Lawmaker* when it is so much more convenient to call Him—God.

What, then, appears to be the reason for the continued reticence on the part of science to officially acknowledge God? The scientist chooses to remain shackled to the "how" technique on the Technological Plateau, and narrowly re-

134

fuses to acknowledge any other intellectual level. He is content to remain in a repressed area, where all he can do is to elucidate the functional processes of nature. With the discovery of $E=mc^2$ the physicist reaches the zenith or climax of his intellectual pursuit. The scientist is confined to the Technological Plateau and he cannot investigate problems on the higher levels where the abstract is judiciously analyzed. Science, therefore, should not offer opinions concerning subject matter that relates to the Theological Plateau, since it is beyond the scrutiny of the scientist. The scientist, by his own decision, remains curtailed in his field of inquiry to the technical functions of nature.

It would be naive, therefore, to expect the scientist to offer valid views on matters that pertain to the Philosophical or Theological levels. If he so desires, the scientist need not speak of God, since science never reaches the plateau whereon such a concept is considered; but this does not imply that the scientist may audaciously deny God. The scientist has no right to express an opinion as to God's being, because the identification of God is not, and never has been, subject to opinion. One cannot foolishly consider God as a candidate for the office of Supreme Being, and thereby believe that it is a human prerogative to cast a vote for or against His existence. To declare that one does or does not believe in God is an absurdity of the modern age. The question of God's existence is not subject to a ballot; nor is it necessary to determine God's existence by virtue of philosophical proof.

God is! Any further verification becomes entirely extraneous and unnecessary.

11

REACHING FOR THE ULTIMATE

Whereas the scientist is hopelessly incarcerated on the Technological Plateau of the "how," it comes within the precinct of Philosophical Man to inquire "why." The term *why* relates to a normative relationship. We may ask "why" elements act in accordance with their distinctly unique manner. Why does each atom behave in accordance with individually endowed features?

The scientist unveils the "how" of table salt when he explains that $2Na + Cl_2 = 2NaCl$. Still to be determined is the "why" of the equation. *Why* does chlorine, a poison, when combined with the element sodium, produce table salt, a useful product? "Why" formulates an ethicomoral query, and searches for an ethicomoral conclusion. If an ethos is discerned, then the universe is not the consequence of an accidental surge. If an ethos is apprehended in nature, then it must imply the forepresence of a Creative Intellect as its sponsor. An ethos that is comprehended by an intellect must definitively be produced by an intellect. The question "why," therefore, inevitably leads to the search for "who." *Who*, indeed, is the Supreme Intellect responsible for an ethos in the universe?

The pity of it is that not very much by way of clarification emerges in response to "who." We can only say God, and then we have exhausted the potential of our mortal intellectual comprehension. As little as we can postulate in the area of the "why," science, by way of contrast, is entirely excluded from any indulgence in such a sphere of speculation. Science can have no opinion in the domain of the "why," because it has restricted its investigation to the operational aspects of the universe on the Technological Plateau. Science is not involved in an examination of the normative relationships of matter, and therefore never specifies an interest in "why." The ethicomoral plateaus are beyond the reach of science.

At times, a presumptuous scientist may commit an act of *hubris* and boldly express personal views concerning the "why," and even the "who"; or to be more precise, he may

136

deny the existence of any "who." When such a   position is enunciated as a doctrine of science, the veracity of its pursuits in many areas becomes tainted by an imbalance of judgment, and this is particularly applicable to the problem of cosmology. The chief difficulty arises in the formulation of the scientific attitude as it approaches the limits of finite cognition.

When science reaches the border of the human capacity for knowing, which is at the threshold of the realm of the "why," it may misalign its logical perspicacity and come to some rather strange conclusions. It may trace a cosmological hypothesis back to the dawn of time, and assume a dense concentration of hydrogen as the ultimate source of the universe. The conclusion that a hydrogen atom or its lesser components may be the *ultimate* should be a somewhat embarrassing confession for the scientist, since such a declaration enshrines the invisible particle as the god of science.

It must be re-emphasized that every individual somehow extols an *ultimate*, which, in effect, thereby becomes his god. Since the cosmology now popular with science ends with the hydrogen atom or its proton, this particle of matter or energy emerges as the *ultimate* source of all formulae, equations, paradigms, and patterns in the universe. The hydrogen atom or proton is thereby espoused as the god of science. Although the deification of a mute particle as the god of science may raise eyebrows elsewhere, scientists, evidently, are hardly concerned with the fact that this invisible charge exhibits no intellectual creative capacity whatsoever for designing a universe.

Certainly, mankind must be eternally grateful to science for providing a marvelous treasure house of advanced technological knowledge. Through the exemplary application of the scientific method, science has mastered the technique of the "how," and has administered it with superlative efficiency. Science knows full well, however, that in its investigations of the technical mechanics of nature, it has succeeded in deciphering only *some* of the mighty riddles of the universe. There remain many more problems to resolve both in the macrocosmic expanse and in the microcosmic domain.

Science does not yet know the source of the vast power generated by quasars, which equals that of fifty million, million suns; science lacks any clear-cut indication of planetary systems around other stars; science cannot yet explain how Hubble's Law for the recession of galaxies departs from linearity at large red shifts; science is still peering into the infinitesimal world of the atom to locate its final minuscule layer; and the list of abstruse perplexities seems to grow ever longer.

Despite admirable advancement on all fronts, and tremendous progress leading to the dawn of the space age for mankind, science must not lose sight of its own limitations. The scientist who assumes that he may winsomely step beyond the confines of the laboratory and casually apply the same scientific principles to resolve intellectual problems on higher plateaus may soon come to realize that know edge beyond the "how" is frustrating and fraught with despair. Indeed, he may very well conclude that it is not possible to apprehend the ethos of the evening star, because man can never directly confront the *ultimate* in a laboratory situation.

Contrary to the conditions that prevail in the research chamber, which is the chief domain of the "how," and where all matter is laid out for convenient, logical analysis, the scene of the "why" resides in an elusive and evasive atmosphere. Man may eventually discover the "how," but he may never know the full "why"; nor will he ever confront the ultimate "who." The best man can do is to become resigned to such a preordained fate.

Let us pause for a moment to interject a parenthetical "what" query. What provokes man, through science, to embark upon the insatiable exploration of nature? Man's motivation betrays a rather ambitious and proud demeanor. Through his research, man eventually hopes to reconstruct the whole cosmic drama. If science would approach its task with a deeper sense of humility and if it would divest itself of its unwarranted hauteur, then science may come to acknowledge that even its investigations in the domain of the "how" fall under the province of a Supreme Creative Intellect. In fact, in each facet of its research, science is examining the visual traces of Godliness.

Science, however, appears to be quite hesitant to identify its investigations as such, and it stubbornly resists any attempt to equate its research with a divine pursuit. Science goes so far as to suggest that the universal process of development in nature should be excluded from any hint of technology, and that such a process reflects the fastidious plan of Darwinian "natural selection," where a Supreme Creator is ignored.

Does the scientist realize that he cannot mollify a "why" query by smothering it with a "how"; nor can he expect to identify the mysterious "who" knocking at the cosmic door, by simply asking "how"? When one finally arrives at the upper echelons of Philosophical and Theological research, "how" betrays its own weakness, because it elicits no information whatsoever. But the scientist remains rooted to the "how" territory and he is limited in speech to a "how" expression. The futility of the "how" when he confronts the "who" fills the scientist with peremptory frustration. The solution, he believes, is to avoid any further allusion to God. Science maintains an aloof attitude toward God. Perhaps it reflects a psychological insecurity and anguish on the part of the scientist when faced with the impossible enigma of a Supreme Being. It is a pathetic situation, because if anyone has reason to extol God, it is the scientist. Laboring as he is in the divine gardens, the scientist, in reality, is a dear friend of God, and not a foe. If only science would recognize this congruous relationship, what a boon it would be for mankind.

12

THE DEVIATED "WHO"

Although man eventually comes to the realization that his quest for the "who" can never be fulfilled, nevertheless, his insatiable curiosity impels him to pursue the enticing, if but elusive, goal. Man proceeds time and again into the fray of this hopeless intellectual challenge. What prompts man to engage in such a Sisyphean task? It is the intuitive conviction that, eventually, by correctly formulating the question "who," man will have gained his finest hour.

If postulating the question "who" marks man's greatest achievement, how does one explain the widespread avoidance of this task by an enormous mixed multitude, who are content to revel in a languid mist of agnosticism and atheism. The human penchant for procacity may be transformed into a mood of apathetic disdain when confronted with the formidable incline rising from the Technological Plateau to the higher levels. As Technological Man approaches the borderline of the finite intellect, he is exanimated by an anguish of hopeless despair. Perversely, however, Technological Man refuses to admit that he has been overwhelmed by dismal failure. In sheer desperation, therefore, he contumeliously constructs an absurd and artificial theory to the effect that there is *no* answer to the question "who." Technological Man thus assumes that he has effectively concealed his utter inability to identify the "who." In denying a "who," he eliminates the thorny problem of the "why," or "wherefore." An elementary corollary to a negation of "who" is the cancellation of the

140

normative "why." Constrained in this manner by the limitations of a "how" pursuit, the cosmological conjectures finally offered by Technological Man appear as inconsistent conglomerations of bowdlerized suppositions.

As a consequence of such a circumvented vision, Technological Man restricts his existential scope to only one other avenue of personal experience—the valley of Biological Man. Bereft of the need for further inquiry, Technological Man cements a solid ceiling above his domain, and provincially maintains that he has reached the pinnacle of human knowledge. Nothing, he pompously declares, exists above his homemade roof. And so, he confines the rest of his days to meaningless gyrations between the self-centered, hedonistic Biological Plain and his cloistered Technological Plateau. Cut off, as he is, from the upper spheres of moral and ethical exposure, the truncated Technological soul readily succumbs to an amoral, idolatrous subsistence.

Technological Man is filled with delusions of grandeur when he assumes that he has effectively dispensed with the "who" query, because the question persistently hovers about like a pesky gadfly. The "who" question refuses to evaporate. A flippant challenge echoes from above, "Who, now indeed, reigns supreme in the castrated cosmology conjured up in the imagination of Technological Man?" The denouement is fraught with ominous shadows. Man has garbed his own finite being in the robe of universal supremacy. He denies existential recognition to any other intellect beyond the sphere of his lowly Technological realm. The inevitable consequence shocks us by its sheer absurdity. Technological Man deifies Technological Man! Man anoints himself as his own god!

An amusing anecdote circulating contemporary scientific circles tells of a research group who finally succeeded in perfecting the ultimate computer. It appeared to be the greatest electronic device ever produced by scientific ingenuity. The complex mechanism was capable of answering any question imaginable. The scientists were overjoyed. They exhibited a feverish pitch of anticipation as they put their technological marvel into operation. At last, they had a computer that knew everything.

One of the exhilarated technicians blurted out a suggestion, "Let's ask it if there is a god!"

The question was duly inserted into the machine: "I-S T-H-E-R-E A G-O-D?"

The computer pompously went through its paces—the room was filled with the terrifying roar of whirring machinery amidst a kaleidoscope of flashing, brilliant lights. Finally, the answer emerged. The impatient and excited scientists grabbed the answer card. The message was brief and simple. It read: *"N-O-W* T-H-E-R-E I-S !"

13

ACCIDENT AND CHANCE

The suggestion that "accident" or "chance" explains the natural evolvement of matter is rather puzzling, since the ultimate source remains unidentified. Nevertheless, it has become the popular view among contemporary cosmologists. They theorize that the universe is the result of an accidental interaction of appropriate substances in matter at a precisely propitious moment. The scientific exploitation of accident dissolves when it confronts the fact of reproduction. An accident does not repeat itself endlessly through a complex process of regeneration. An accident suggests an unusual deviation, which occurs as a peculiar coincidence of circumstances, and which should hardly expect to be repeated.

Nature does not abide by accident, but it does operate through chance; not, however, in accordance with the current scientific interpretation of the term. Science considers "accident" and "chance" as handmaidens of creation; and that the entire universe is not an emanation of design, but of sheer purposeless coincidence. A more discreet analysis suggests that the time/spatial prospect of chance in nature affects the particular, but never the universal.

The procreative urge in nature produces an abundance of variations. Protons eventually become molecules; plants shed an abundance of seeds; fish lay many eggs; and mammals multiply in munificence. Nature is exceedingly generous in providing the essential material and an incontrovertible pattern for generation and regeneration. An entire planet develops and perpetuates a life cycle, despite the delicate, gossamerlike balance it must establish in an ecological sense. The mathematical probability of a planet like our own earth emerging strictly as a consequence of pure "chance" is very dim indeed. Our planet developed, more likely, because as a cosmic seed, it fortunately fell into the orbit of the sun's "zone of life."

"Chance" plays a significant role in the appearance of the particular. Among the multitude of acorns that are shed by the oak tree, few, if any take root to produce another tree. When, however, such an event does come to pass, the particular acorn that is privileged to sprout anew is the consequence of a time/spatial coincidence. It may be said that "chance" fostered the growth of that *particular* oak tree. Consider for a moment our own personal existence. If our mothers would have conceived at a different moment of time, it is quite probable that we would never have seen the light of day, but that privilege would have been accorded to a brother or a sister. Our appearance as a *particular* ego specimen, so to speak, is a fortunate consequence of "chance."

In the area of design, however, no element of chance prevails. The result is definitively ordained beforehand. Our parents had to give birth to human offspring. The acorn does not blossom into an apple tree; nor does the rose seed emerge as a lilac bush; nor can a dog bring forth a litter of kittens; nor will the salmon's eggs sprout into fledgling robins; nor is the proton capable of behavior other than that with which it has been endowed. The *universal* is preordained, and its perpetuation is no accident. It is not only the predesignated paradigm for a desired model, but it also accommodates a multitude of variable deviations that may accrue in the event of unanticipated conditions. Even erratic abnormalities that result from molecular or cellular aberrations are preordained. In fact, the startling grotesque freaks produced by man-made or natural mutants are also predetermined formulations.

The preordination of behavioral function may apply to a planet as well as to a proton. In fact, why shall we not consider a planet to be a fertilized cosmic seed? When it orbits in the favorable "zone of life" around a star, it must germinate and develop a vitalized, variegated inorganic and organic community of substance in consonance with a fixed plan of planetary growth and behavior. The planetary seed may be impregnated with as preordained a design for its eventual development as any other prescribed incipient force in nature. Given the opportunity and the proper prevailing conditions, matter will consistently conform and behave in accordance with its formulary structure. Its behavior is not subject to any chance; nor is its innate paradigm the result of any accident.

Was it an accident that life appeared on planet Earth, and not on Venus or Mars? When the cosmic seeds were sown throughout the universe, the resultant planetary bodies would not all emerge as hospitable homes for the development of life under their suns. In our solar system, Earth alone may have had the good fortune to fall into the orbital "zone of life" around our modest star. It, therefore, was enabled to fulfill all the coincidental time/spatial requirements that bring forth a lush and vibrant planet. Perhaps in the orbital systems of other distant stars, the counterparts

144

of our own planetary likeness are presently swirling about in a similar organic manifestation wondering if planets accompany our sun. Perhaps they may be older than we are, and they may already know that there is a planet "Number 3" in the throes of early Technological awakening, but menacing its own survival by promoting a stunted Biological infancy as the subsistence for its civilization.

In any of these inhabited planets, chance applies not to the nature of their appearance, but to the time/spatial coincidence of their appearance. If Venus would have fallen into the Solar System's "zone of life," then the story of a human civilization would have been recorded on our sister planet, rather than our own. Like the acorn that blossoms into the inevitable oak, so does the cosmic seed indubitably develop into a preordained, life-supporting planet, provided that, like the *particular* acorn, the planet swings into a propitious environment for germinating and flowering its life potential.

Of course, the factors for planetary flourishing are much more complex than those for the acorn. Nevertheless, a parallel does exist in that they both reflect the principle of coordinating their functions in accordance with a grand master plan in nature, which provides for the pullulation of a universe.

The question "who" must now be formulated. "Who" imparted the dynamic cosmological pattern unto nature? "Who" postulated the formulae to which all matter is pledged? "Who" sponsored and brought forth the first atomic particle? "Who" set in motion the design for growth, expansion, condensation, and reproduction? The answer is certainly not man; nor does "Nature" slake our intellectual curiosity. Only one explanation assumes rational validity. The "Who" is God and none other!

In God's universe, the certitude of the *universal* is without peer. There can be no deviation from the divinely ordained paradigm imposed upon matter. If one does not see God in the gemmation of the acorn or the cosmic seed, it does not mean that God is not there. The power of God *is* there, and it is discernible throughout the enormously complex blueprint of nature. In this respect, our diminutive globe fulfills its role in accordance with the divine scheme.

With the appropriate reconstruction of his ultimate objectives, Technological Man may conclude that the final revelation of the genesis source is unattainable. The mysteries of the universe become more profound, even as science becomes more knowledgeable. The incongruity of being cannot be explained by Technological Man. There are advanced mathematical equations that have no answer. In his microcosmic investigations, the research scientist has not yet reached the rock-bottom layer of the atom. It is now theorized that the infinitesimal protons and neutrons may be composed of a smaller entity, and recently, these have been dubbed with the somewhat facetious name "quarks."

Physicists have presently identified four types of quarks in the microcosmos, which they have quaintly designated "up," "down," "strange," and "charmed." These pursue a variety of triad combinations to formulate the particle structure of protons and neutrons. But this is not yet the end of the story. Each quark is comprised of three smaller quantities that are called "colors," and this still may not be the ultimate minutiae of matter! Nevertheless, the triad formulation of quarks and their "colors" is an impressive demonstration of nature's persistent adherence to relational symmetry.

Among men of science, there are those who declare that a total understanding of the forces governing particle behavior is impossible. Despite this conclusion, Technological Man continues his search for the perfect symmetry and ultimate simplicity that he believes must reside at the core of all nature. A force that promotes such singularity and yet

146

harmonious extension calls to mind the concept of God more than anything else.

In the macrocosmic universe, the puzzling phenomena of quasars baffle the world's leading astronomers. When the astronomical union met in Hamburg in 1964, only nine of these blue, ultradistant points of light were known. By 1967, some one hundred of these "stars" were identified. They were considered to be farther away than any other visible objects in the sky. In 1974, more than 250 quasars had been located and catalogued. Since then, their number has increased, and although their precise nature and distance from our solar system remains a mystery, recently, a pair of American astronomers declared them to be the exploding nucleus of a distant galaxy.

It seems to me that the human mind all too often suffers from a narrow constriction in its thinking because of its naturally finite composition. As a result, man's hypotheses concerning outer space has been severely circumscribed commencing with the earliest records of cosmic speculation. The universe of the ancients was smaller than the universe of the medieval scholars; and their views of the world were more confining than those of the early scientists. The trend of conceiving cramped cosmologies continued into the twentieth century. Only some fifty years ago, astronomers regarded the mysterious galaxies as nothing more than single stars exploding in nearby space. Today, the galaxy is identified as a distant stellar metropolis comprising millions of stars.

Science should not be criticized for its cautious approach, and perhaps its circumspect method generated the current narrow hypothesis offered in connection with the unexplained quasars. In view of the long, past record of underestimating nature's grandeur, why should not one boldly conjecture that quasars are more than the exploding nucleus of a galaxy? The nature of these quasistellar phenomena

147

may one day be known. Until their enigmatic qualities are discerned more definitively, they remain a subject for wide speculation, some of which may be fanciful and some of which may be intuitive. In this latter vein, I should like to offer an element of theoretical interpretation to what has already been popularly assumed.

Despite its minute appearance, the quasar is an exceedingly luminous stellar object; in fact, quasars appear to be brighter than the largest galaxies, although they are much smaller in our sighting instruments. More unusual is the amount of energy that they radiate into space. For example, the enormously bright quasar 3C 273 emits a thousand times more energy than our own Milky Way galaxy, which comprises some 100,000 million stars, yet in its compact size, 3C 273 seems to be but a trivial fraction of our galactic volume. Scientists are hard put to explain the anomalous phenomenon of such a small stellar object that produces such a vast output of energy.

Among other theories, the quasar has been considered to be a collection of millions of stars compressed into a tiny compact volume, wherein regular supernova explosions emit tremendous forces of energy. Some astronomers consider that the quasar's enormous energy derives from the decimating consequences that occur when matter and anti-matter meet in a direct confrontation.

The crux of the problem hinges upon a convincing determination of the distance between the quasars and our own point of observation. For this purpose, measurement of the light from the quasars denoting the red shift toward the longer wavelength of the spectrum has been somewhat dubiously, if not decidedly, accepted as a means of denoting its place in the universe. The enormous red shift indicated in the light of the quasars suggests that they may lie at tremendous distances from the planet earth, and possibly at the very outposts of the universe itself. On the basis of

148

such computations, the nearest quasar has been estimated to repose some two thousand million light years from our solar system, and the remaining quasars are still farther out, with some residing even beyond the zenith of our observable universe.

It is considered that the present expansion of our own universe may have commenced approximately ten thousand million years ago, and our Milky Way galaxy is in the neighborhood of five thousand million years old. If such be the case, and we are now viewing in the quasar appearance cosmic phenomena as it was some two thousand million years ago, or more, then we are gazing upon quite an early vestige of universal structure.

Under such conditions, I am tempted to assume that the quasars are composed not of millions of stars compressed into a very small volume, but of clusters of galaxies at an early stage of cosmic spawning, prior to their dispersion into the vast reaches of space. In other words, what I am suggesting is that these supradistant agglomerations of stellar phenomena, emitting incredible amounts of energy in light and radio waves, constitute an entire urban universe composed of clusters of galaxies at an early stage of their development and expansion.

Thus, contrary to the notion that our universe is the one and only extraordinary cosmic island of stellar matter, why may we not assume that many more island-universes exist throughout a whole suprauniverse? Perhaps, like the ancients, our concept of matter remains narrowly compressed in relation to the more realistic condition of the expanse. It may turn out that the universe of the twentieth century is only a minor fraction of a mammoth suprauniverse. If it is theorized that each quasar is a separate multigalactic universe, then our own universe of galaxies diminishes in significance, as does our own galaxy, the Milky Way. What is accomplished by postulating that our own universe may be but a minuscule quasar-speck in a supra-universe composed of billions of quasar multigalactic universes? Man becomes smaller, and therefore, should be more humble; but, God becomes greater, and therefore, more praiseworthy.

Intellectual Man, however, is now sorely puzzled. It appears that his singular position in the universe is not as extraordinary as he considered it to be at the outset. Suddenly, it dawns upon him that he may be but an infinitesimal speck in the cosmos, whereupon his once august demeanor of total centrality crumbles. To learn that man is not the center of creation is a shattering blow to human vanity.

If man is but an obscure particle amidst a mind-staggering expanse of matter, and if his individual welfare evokes no great concern on the part of a stern-faced nature that spawned him, then why was he endowed at all with an intellectual capacity to perceive and comprehend such a state of affairs? Why should he be granted the rational ability to detect the fact that, in the eyes of Mother Nature, his particular existence commands no special consideration? The perfunctory role of nature makes it imperative that we delve beyond the domain of the "wherefore."

14

THE PLIGHT OF THE PARTICULAR

A very sagacious cocoon for emergent life was spun on Planet Earth by Mother Nature; however, her main concern seems to be the fostering of species, and ensuring their perpetuation. Nature patently ignores the trials and tribulations of the particular. Nature is clement toward her family, but obdurate toward its individual members. Small wonder, then, that an analysis of the human situation reveals wholesale anguish and anxiety as the forerunners of despair and hopelessness.

Such forlorn attitudes are quite prevalent in contemporary

society. The atheist-existentialist, for example, moans about the vacuity of human existence. There is no logical explanation for the inanity of man's being. In fact, it appears that nature played a cruel hoax on its supreme product, man. It endowed man with an intellect, with which he was enabled to perceive the utter futility of his own essence.

It is not easy to suffer the awareness of a meaningless existence. No organic specimen, other than man, has been inflicted with such a prepotent imposition. It hardly inspires a zest for life when one observes nature smiling broadly upon the species as an entity, and favoring the biologically superior product, while brusquely turning her back upon the tortured human psyche. Why was man chastised by being granted a brain that may apprehend all this?.

Has man's superior mental apparatus brought him greater security or happiness? An ego-awareness faculty, such as man possesses, becomes a hollow mockery in face of the emotional torment it causes him to endure. His intellect weighs upon him like a burdensome albatross. It would have been a far better thing for man to have been deprived of such a keen, discerning thinking-equipage, because as a simple biological creature, he would never have comprehended the sordid plight of mortal existence. If man were nothing more than a menial biped, he would have been spared the voluminous psychological stress and emotional sorrow to which he is presently subject.

Now that man, through his superior intellect, has unlocked a Pandora's box revealing the gross human condition in all of its misery, to whom may the pitiful individual turn for comfort, solace, and sympathy in the dark hours of isolated suffering and loneliness? Shall he call upon Mother Nature? But Mother Nature could not care less about his pathetic predicament. If the distressed subject persistently squints at the world through a constricted, agnostically dominated vision, then, indeed, there is no one to turn to but himself.

151

Life, as seen by the atheist-existentialist, is an absurd carousel. It comes from nowhere and it goes nowhere, all to the accompaniment of a garish musical clatter. Existence is a kaleidoscope of crazy-quilt patterns; and because it makes no sense, *being* may be equated with *nothingness*.

Accordingly, the atheist should not misrepresent the fact that in the eyes of Mother Nature, the plight of the particular evokes little sympathy. Once nature's callous demeanor toward the particular is acknowledged, the atheist becomes hard-pressed to offer a sensible explanation for the evolution of a unique human ego. If man, as a biological product, is nothing more than the chance eruption of a peculiar specimen, whose only function as a particular is to interlink another element of the human concatenation for the simple perpetuation of an irresolute species, then what is the purpose of endowing him with a faculty for self-conscious awareness?

This question becomes an especially thorny issue for the Darwinians and contemporary behaviorists, who conceive the cosmic cycle of nature as operating through a doctrine of "natural selection," so that nature only produces that which is required for the survival of the species. If such is the case, the highly developed brain of species *Homo sapiens* serves no valid purpose. In fact, in view of his marked penchant for fraticidal demolition and global suicidal behavior, modern man represents a blatant antithesis to the law of "natural selection."

There was no biological compulsion on the part of the evolutionary process to produce a rationally minded, ego-conscious, and self-analytical specimen. It would have been more within the operational design and function of nature to generate an emotionally sterile and rationally supine creature bearing the same image as man. Such an anthropomorphous being could adequately populate a planet through the instincive procreation of a subintelligent, but psychologically tranquil, species, *"Homo simpleton."* A species *"Homo simpleton"* would have been the logical Darwinian progression beyond the higher primates, and endowed with an appropriate set of animal instincts necessary to conform with the survival pattern of nature, it would have made a logical capstone for the evolutionary process. Not *Homo-sapiens*, but *"Homo-simpleton"* should have populated this planet, because nothing more than an exceedingly shallow-minded folk should have appeared at the end of the long, Darwinian chain.

The facts of life, however, are otherwise. Modern man emerges as a paramount prodigy of the extensive cosmic scheme of nature, and this poses a baffling situation. In the eyes of the atheist-existentialist, man's appearance is inexplicable; in the sight of the honest scientist, he is an anomaly. Furthermore, the human acquisition of a high-grade intellect discredits the theoretical structure dogmatically developed by the Darwinians and the latter-day behaviorists.

Why should a universe emerging through the rigorous throes of pure accident and sheer chance evolve an intelligent specimen of matter who has the ability to analytically contemplate his own ego structure? As a product of mere coincidence, the formulation of an advanced, rationally equipped entity makes little sense. Even from an orthodox pantheist's point of view, the chief purview of nature suggests a desire to activate a biologically thriving universe, which requires nothing more elaborate than a variety of creatures endowed with instinctive awareness for survival and propagation. An emotionally turbulent, psychologically complex, and intellectually superior human species should never have appeared on a cosmic menu, if it is assumed that the Master Chef is nonexistent. Under such conditions, man is not only unnecessary, but in contrast to rationally deficient and mechanically or instinctively motivated matter, he appears to be an incongruous absurdity.

The absolute emergence of rational man, therefore, indicates that the cosmos is not the consequence of an accidental cataclysmic catastrophe. This conclusion, however, does not yet explain the circumstance of the disaffected particular. The impersonal, mechanistic disposition of nature toward the individual embodiment of a species remains a disturbing enigma. Why does nature hardly exhibit any concern for the destiny of the *specific* particular? Her only interest is to preserve a particular so that it may service and perpetuate the species.

Nature's lack of interest in the particular is quite evident in the area of propagation. Nature disperses its seeds in such generous proportions that one may chide her for exercising impetuous wastefulness. For every seed that takes root or that germinates, there are many dozens, hundreds, and thousands that are scattered uselessly, and their fruition never sees the light of day. Furthermore, man is appalled at nature's indiscriminate exercise of her powers. Her destructive fury may strike forth against the wicked and the innocent alike. Why, indeed, does Mother Nature display a heartless amoral attitude toward the struggling particular?

This brings us to a vexing dilemma that determines the entire outcome of man's personal relationship with God. Since the Divine Personality radiates its message through nature, does the lack of biological concern for the welfare of the particular, as found in the cosmic scheme of nature, reflect the divine will of God? If it does indicate the divine attitude, then the abyss between man and God can never be bridged, and the human endeavor to address the Supreme Being is hopelessly futile.

An understanding of man's position as a particular must be two-fold. On the one hand, man is nought but an innocuous particle of nature, and in this respect his particular identity is of slight consequence. However, man is also something else; but this supernumerary status depends upon the point of view man, himself, assumes toward the cosmos.

If man decides that he will occupy only the biological lowlands where the law of the jungle prevails, then his particular existence fades in importance and diminishes in its specific significance. Neither Brute Biological Man, nor Domesticated Biological Man, nor Civilized Biological Man can file a claim for meaningful existence, because they choose to identify solely with the base biological thrust of life, where the particular survives in an impersonal atmosphere of dispassionate instinctive awareness.

In the valley of selfish biological survival, man fails to perceive a higher ego identity. It is when man arrives on the Theological Plateau, and certainly on the Prophetic Summit, that a distinguished ego consciousness comes into view. Lower Biological Man, however, is so busily engaged with matters of the flesh that he ignores the opportunities to reach the higher spheres in life. Resolutely, he prefers to occupy his sordid environment, where the chief occupation is to satiate crude biological appetites and, principally, the sex drive. Lower Biological Man swings with nature-in-the-raw, and in so doing, he slips to the nadir of existence. As a particular in that realm, he commands an insignificant status.

Despite its shortcomings, the domain of Lower Biological Man appears to be exceedingly popular. It attracts vast numbers of the global population to its banner, and they render it homage fit for a deity. Formal philosophical schools of thought have been devised to accommodate the mass penchant for a life-style extolling the biological motif above all else. Atheist-existentialism stands out as a notable example.

The atheist-existentialist relates the particular to the *en soi* category of biological instinctive awareness. For this reason, the atheist-existentialist deems life to be a futile struggle. Human existence is confined to the biological valley of the three lower biological specimens: Brute Bio-

155

logical Man, Domesticated Biological Man, and Civilized Biological Man. The upright two-legged creature is conceived as nothing more than an educable beast. Man must struggle along for survival with all other instinctively aware creatures in nature. As a consequence, despair, anguish, and futility prevail.

In a broader sense, man is more than an instinctively aware creature. Man has developed the unique ability for self-conscious cognition. With the inimitable capacity to identify his own ego, man advances to a higher status of existence. He becomes superior to the common category of organisms who thrive through instinctive awareness. Simply stated, man, alone, knows that he has an ego. Only man is cognizant of an ego within himself and within his fellowman. An important facet of ego-cognition is the aptitude to perceive that, within a fellow human being, there also resides the power for ego-cognition.

The intellectual expansion of man through his power for ego-cognition inspires within him an intrepid resoluteness to identify his own particular being as an essence of vital significance in the universe. His self-knowing skill convinces the individual that he is endowed with a sense of special importance and purposefulness.

If the outlook of man is a criterion, then an ego-cognizable specimen finds existence unbearable unless the particular is invested with meaning. If no purpose is evident for the creature blessed with ego-cognition, then life becomes a harrowing drama of grievous anguish and wearisome despair. To the atheist-existentialist, who nullifies all *raison d'être,* existence becomes a burdensome exercise in futility. Man finds it difficult to endure a situation whereby his particular essence is merely meshed into the survival pattern of non essential flotsam floating along a stream of cosmic residue. Man cannot abide the existence of an *en soi* if life is to convey even the slightest apparition of meaning.

156

Unfortunately, the atheist-existentialist, in directing man toward a *pour soi* essence, fails ignominiously to provide the human specimen with a worthwhile prescription for achieving a valid goal in life. The *pour soi* directive of the atheist-existentialist is as empty, vapid, and nugatory as is his otiose *en soi*.

A creature possessing the potential for ego-cognition represents an extraordinary breakthrough in the normal functional pattern of nature. Heretofore, all existence was either inorganically automated, or if organic, instinctively motivated. One could hardly anticipate the emergence of a human specimen with the power to rationally analyze and influence nature's organic and inorganic operation.

Man is indeed an anomaly. In one sense he is a biological product with hardly any redeeming features to warrant special notice. In another sense, however, he stands squarely above all nature as a regal and supremely intellectual giant. As a percipient *persona*, man cannot sustain a limited state of simple instinctual awareness. He feels impelled to generate an essence that may transform his existence into a meaningful adventure.

Although the psychological essence of human ego-cognition is biologically dependent, it nevertheless encourages man to assume that he is the proprietor of a unique being, and therefore, he is vastly superior to the diverse miscellany of nature. Man asserts that his particular essence does qualify for a distinctively respectable status of incomparable significance. Because he is equipped with such a highly proficient intellectual talent, man makes bold to declare that his particular being arises as a purposeful event in cosmological history. In fact, man knows that he does relate meaningfully to the stars, the galaxies, and the universe.

Ultimately, it dawns upon man that the special endowment of a profound intellectual capability is not merely a fatuous license for his own sophisticated, biological amusement; rather it is an imperious mandate charging him with the obligation to search beyond nature for the source of all things. Man's ability to postulate the query "wherefore" imposes upon him an imminent responsibility to seek the "Who." The process of this search for the Grand Master is marked by a somewhat devious route. Man must first embark upon a penetrating, introspective ego analysis before he may acquire a discerning revelation of the universal absolute.

The circuit to the Divine Personality commences with the unfolding of the mortal "I" ego. Through ego-cognition, man may assume that he holds a more valid status in the cosmic identity. His existence exudes the quality of a vital essence. His particular ego occupies an eminent category in the unfolding destiny of the universe, because man *knows* that he relates historically to the first proton. By reason of this primary awareness, he recognizes that he shares in some small fashion in the determination of cosmic eschatology. Man discovers a unique relevance in tracing the source of his own being. By further extension of the faculty for self-conscious cognition, man is able to discern a meaningful premise for his own identity.

How does man proceed to certify his endorsement as a meaningful particular in the cosmos? He records his biography from generation to generation, and he ratifies his book of existence as an important volume in universal eschatology. In fact, he believes that each individual record in his encyclopedia of life is an essential text in cosmic history.

Man concludes that his existence does leave a mark upon the world. He is convinced that the consequences of his own particular adventure stamp an indelible imprint on the destiny of the cosmos. His being is a microcosmos reflecting the macrocosmos. His life is not a mere shadow flitting from insignificance to nothingness, because through the remarkable identification of his I/mortal ego, he can postulate the divination of God. Indeed, through the awareness of his own "I" ego, he can infer the presence of a Divine "I" ego. Authentic existence is thereby within the reach of man. To qualify for authentic existence, the self-conscious cognition faculty expands its comprehension of the particular "I" ego, and eventually, it becomes possible for man to conceive God as an "I" Divine.

Man's particular being, therefore, is vitally significant, and the diminution of his "I" ego through his passing is a tragic event. Moreover, if man ever becomes extinct as a result of his own malevolent disposition, no front-page headline type will ever be large enough to justifiably herald such a horrendous catastrophe. Of course, no typesetters will have remained, anyhow, to record such a disastrous event.

It is therefore reasonable for man to consider the emergence of his ego-cognition faculty as an indication that such a largess is a special grant from a power beyond the instinctual, automated operational design that is so clearly manifest in nature. There is no doubt but that the species *Homo sapiens* arises as an unusual phenomenon in the universe. Here is a strange particle of matter of the cosmos that developed the skill to comprehend its environment; the ability to contemplate itself; and the proficiency to search for its source in time and space. Could such a being be the quite accidental consequence of an impersonal, natural occurrence? The rare thinking organism could only conclude otherwise. Man's appearance had to be the divinely inspired and divinely ordained consummation of a cosmic scheme.

Rational man refuses to credit a blind, coincidental interaction of impercipient atoms with the competence to formulate its own paradigm for the emergence of a specimen able to contemplate itself. The appearance of a self-conscious awareness factor in nature must have been conceived

159

through the grace of a higher intellect. On the entire horizon of nature within the periphery accessible for human analysis, man alone stands out as a specimen who may study his "I" ego. No other organism known to man can recite the apparently simple first-person pronoun, "I," and comprehend its significance.

The special human capacity for ego contemplation opens the intellectual gateway for the postulation of a Divine Personality iridescently resplendent within the sacred folds of its own Supreme "I" Ego. In taking this bold step forward toward a more personalistic identification of an Eternal, Immutable God, man presumes to impart a distinctive and meaningful quality to his own *particular* being. No other creature exhibits a personalistic analysis of its own ego. No other living specimen known to man discerns the concept of God, let alone a Divine Ego. The exclusivity of this profound eruditeness achieved by man is reason enough for him to assume a superior posture of purposefulness.

With such a keen analytic aptitude, man rises above the vortex of nugacious biological existence and invests his individual appearance with meaning. He refuses to accept the nondescript designation that Mother Nature customarily imposes upon her offspring. Man substantiates the claim for individual pertinence by endowing each particular with its own characteristic name. The particular's importance and singularity are confirmed by assigning each individual member with a personalized nomenclature. This practice certifies the distinctive identity of the I/ego. In a similar vein, man assumes that God must also be known by name, so that His all-meaningful being may remain supreme in the eyes of His creatures. Subsequently, the custom of name-calling is applied to the rest of organic and inorganic matter, but this measure is undertaken by man merely for the sake of establishing a convenient means of description and identification by which they, too, become part of a universe of meaning.

The fact remains that man, alone, communicates through the process of names. Man, alone, resists nature's impersonal attitude toward the individual specimen. Man defies nature's callous disposition toward the particular by providing the *persona* with purposeful significance through a historically applicable relationship involving the cosmos and the Creator.

Man may take pride in the fact that his unique cognitive ability is withheld from every other earthly, organic creature. The singular manifestation of ego-cognition accorded to man, therefore, marks him as a distinguished and purposeful existent being in nature.

A more profound revelation of ego-cognition occurs with the enunciation of the divine Decalogue's first commandment: "I am the Lord, thy God. . . ." God is identified as exhibiting a self-conscious awareness faculty. Inasmuch as self-conscious awareness is evident only in the intellect of man, its parallel source must be located in a superior realm, namely, in the domain of the divine. It is therefore considered to be a divine gift to be able to enunciate the word *I*. Man has been so blessed, only because he reflects the image of the Divine Personality. The revelation of a divine ethos in the universe becomes possible when God's identity is formulated and expressed as a Supreme "I." The I/mortal may then relate to the I/Divine, following which, man may proceed to perfect his intellectual prowess so that he may discern the universal moral directive within the design of a divinely inspired universe.

The unfoldment, then, of the I/mortal leads to an apocalyptic unveiling of the I/Divine. When man discovers that he is an "I," he feels impelled to adduce a similar transference of ego manifestation to the Divine Personality. If man displays an "I," then God must surely exhibit the

161

Supreme "I," and it is through the omnipresence of the Supreme "I" that man may experience a divine revelation. Thus, the divulgence of the I/Divine enables man to perceive the will of God in the universe.

The quintessence of authentic existence resides in the essence of divinity. God's being is the purest and most perfect form of existence. If man could reflect some aspect of the Divine Being, he possibly may acquire a modicum of authenticity for his own mortal being. If man gains even a whit of authenticity in his life, he may justly claim a token of special respect and meaning for his particular identity.

In what manner, then, may man demand to be distinguished from the ordinary existence of a commonplace substance whose particular is of little interest to Mother Nature? How may man garner the appropriate esteem so that he may claim a meaningful measure of relevance for each single human specimen? Such a role is fulfilled when mortal ego-cognition basks in the resplendency of the I/Divine. Since the I/Divine formulates the basis of all authentic existence, the I/mortal, by emulating the divine essence of ego-cognition, justifiably may conclude that in the Divine-mortal ego-similitude, human existence acquires authenticity. When man, in enunciating "I," perceives that he is afforded the awareness of his own humble mortal ego only by virtue of the beneficent grace of the supremely Divine "I," then he may infer the unfoldment of such similarities in personality structure as being a coadunate concomitance of authentic existence.

God emerges from His role as the transcendent composer of the glorious cosmic epic, and reveals Himself as the supreme I/Divine, whose particular essence manifests a unique singularity and individuality. By emulating the ego-cognition evinced by the I/Divine, the I/mortal, although trivial by comparison, nevertheless, through the privilege of *imitatio deus*, assumes for its own mortal particular a singular quality of unique identity and significance in the cosmic drama of the universe.

As indicated, an interpretive analysis of the self-conscious awareness faculty encourages man to conceive himself as bearing an I/mortal ego that is a minute reflection of the I/Divine. Man should not be too harshly criticized if, subsequently, he boastfully announces, as is recorded in an early Genesis passage, that he was made "in the image of God." Indeed, if both man and God exhibit ego-cognition, then an imagery parallel between them is not too facetious a pronouncement for the ancient biblical scribe.

A sharp contrast becomes evident as life casts its flickering shadows across the screen of existential reality. Whereas in the broad sweep of nature the particular is of little concern, on the Theological Plateau and on the Prophetic Summit of man's domain the particular is never regarded as blithely expendable. Nature exercises a cold, impartial attitude toward the particular. The single specimen is of interest only as it serves to sustain the propagation and continued survival of the species. On the higher plateaus of man, however, the particular enters into a covenental relationship with God. The particular speaks to God; the particular considers the mysterious personality of God; the particular ponders on the problem of life's purpose, and therefore, searches assiduously for the divine message. This rapport and communion that the I/mortal endeavors to establish with the I/Divine helps to convince man that he is granted the privilege of considering his ego-being as engendering a meaningful significance. Man cannot help but conclude that he is a vital link in the chain of cosmic destiny as outlined by the recondite master plan of the Supreme Being.

The enlightenment that man derives on the upper intellectual plateaus in relation to the identity and aspirations of the ego, fills him with a true sense of worth. He deems it valid to pursue the enigma of life, and in this metempirical search, he hopes to gain authentic existence. Man is convinced that authentic existence is within his reach. Where, however, does it seem most propitious to encompass it?

The lower Biological mode of subsistence offers small hope for capturing the authentic in life. Little wonder, then, that the particular on the lower Biological levels is so ruthlessly ignored and so harshly disdained. One must climb, therefore, to the intellectual plateaus, and particularly to the Theological and Prophetic heights, if one is to discern authenticity. In the rarefied atmosphere of the upper Intellectual realm, the distinctive recognition of the I/mortal ego, and its relation to the revelation of the I/Divine, invests man with a noble posture for living.

As man descends from the upper Intellectual spheres along the incline leading to the Majestic Biological Plain, he feels that his life is filled with a sense of meaning; his singular appearance is important; and his *particular* identity has purposeful value. A basic element in man's search for authentic existence, therefore, is the appropriate identification of the I/mortal ego, so that it may fulfill its function in the establishment of a meaningful relationship with the I/Divine. The I/mortal must strive to reach the plateau that is most promising for a rendezvous with the I/Divine.

In a frantic desire to reach God, man has applied his creative talents to the maximum, hoping thereby to arrive at the proper formula. As a consequence, a plethora of ritualistic doctrines were developed over the centuries, and these were designed for the purpose of making available to man a prescription for divine communication. The rituals established in the course of many generations as a means of reaching God now possess the distinguished characteristic of tradition. In addition to his many other traits, man is a decidedly sentimental creature, and an identification with tradition, therefore, promotes a more favorable psychological climate for intrepid living.

Assuredly, both ritual and liturgy are important elements in man's search for authentic existence. They are essential as expressions of his determination to reject the stigma of expendability that nature stamps upon the individual specimen. By pridefully addressing God, and this is usually accomplished through established ritual and liturgy, man asserts that he is *not* expendable. He declares in stentorian tones that his *particular* ego identity is not a passing fancy in the wind. His essence, although mortal, has value, significance, and purpose.

Man, however, recognizes that he is weak, both physically and psychologically. He therefore requires the gracious guidance of a higher power to aid him in securing a significant essence. In order to win a seal of meaningful approval for his own humble status, man suppliantly turns heavenward and implores that his own "I" ego—his own I/mortal—may relate in parallel conjunction to the I/Divine. Just as the I/Divine is unique and supremely indispensable in its particular identity, so does man beseech that his mortal "I" may reflect a spark of this uniqueness, a spark of this indispensability—in short, a spark of meaningfulness in the dispassionate cosmic panorama of the universe.

The classic outcry of the Psalmist, "Bless my soul, O God!," well reflects this notion. It suggests through implication: "Exhibit Thy awareness of my essence; of my ego-identity; because only through my relationship with Thee, O God, does my life assume extraordinary value. Grant unto me the privilege of being acknowledged as a significant particular. Do not cast me into the melting-pot of estranged biological phenomena. Do not relegate me unto the meaningless maelstrom of lower biological subsistence, where the particular is measured only in terms of its pragmatic biological efficiency. I implore Thee, O God, bless me with a more meaningful cognizance of Thy Being, so that my own being may thereby acquire a greater degree of authenticity."

Thus, by pleading for a demeanor of worthiness, man hopes to find the way to perfect his own existence so that it may be authentically developed. The alternative is to float aimlessly along the broad, impersonal stream of purportless, particular matter scattered throughout the universe. Man, however, is impelled to presume that his being is different, because he is the only identifiable specimen possessing the rare quality of ego self-cognition.

Man's *particular* identity, therefore, is important. His unique essence has significance, and somehow, man concludes, his being does affect the destiny of the cosmic drama. Man interprets his appearance as the crowning-point of nature; and why should he not assume such a regal stature? Consider, in all of creation to which man has access, there appears no other creature endowed with the unique capacity to conceive through a self-conscious awareness of a mystifying, mortal personality, a glimmer of the ineffable I/Divine. As a fitting climax to such a profound revelation, may not man boldly presume to endow the whole universe with meaning?

Therefore, man should not be branded as being overly arrogant when he concludes that nature produced a whole cosmic panorama of matter, so that one of its products may look in on itself; analyze its own condition; declare its independence from the apparent absurdity of the *en soi*; identify itself by name; perceive the Divine Being; and furnish its own particular entity with meaning by relating in a personalistic manner, through confrontation and dialogue, to God.

The problems that arise at this point become quite profound and abstruse. How near can man come unto God?

How much of the Divine Personality can man truly discern? How involved a dialogue can the human *persona* develop? The analysis of these enigmas calls for the most concentrated application of human rational perspicuity.

In order to comprehend the futility of any mission, it is necessary to attain a high level of knowledge in that specific field of endeavor. The incongruity of advanced mathematical equations that have no solution can best be understood by one who has achieved the astute sagacity of an expert mathematician. The futility of the quest to discern the ultimate secrets of nature that lie beyond the reach of the human finite mind can best be appreciated by those who explore the domains of the higher plateaus.

Technological Man gains a modicum of awareness of the formulation of nature through a patient, persistent, and thoroughgoing analysis of the universe of relational knowledge. He finally learns "how" nature operates. Although he does not learn the "why" of things, because he lacks the inspiration to know the "who," Technological Man nevertheless makes an extremely valuable contribution to the storehouse of human knowledge by developing man's ability to compose the necessary question—to arrange the puzzling equation that has no answers.

In his "how" research, Technological Man peers into the complicated strata of nature, only to discover that the configurations and the aspects of relational knowledge eschew any semblance of personality. Within the limited accommodations of a "how" investigation, Technological Man may find an "it," but never a "who."

Philosophical Man, in his search for "why," finds that the answer points to the mysterious "who," and identifies this being as "He"—the hidden Divine Personality—or more appropriately, for Philosophical Man, the Supreme Being. Technological Man never confronts a personality, and therefore, he can never perceive a "He." It is a concept

167

beyond the most advanced cognitive faculty with which Technological Man is concerned on his domain.

Theological Man, searching for a Divine Thou, soon learns that man can never attain the answer to the ultimate "who." At best, all that Theological Man can hope to achieve is a proper phrasing of the question. The essential significance of the Personality Mysterium is beyond human comprehension.

In view of the fact that his quest is so hopeless, shall Theological Man succumb to a mood of despair? Decidedly not, because a knowledge of certain aspects of the personality of the Supreme Creative Intellect may be derived from finite nature. The perception of the glittering rays of the Divine Intellect sparkling within the bosom of nature like precious gems is reward enough for the soulful human rational faculty.

Although the Master of the Palace seems to be beyond the pale of immediate recognition, nevertheless, the aura of His Divine Presence pervades its stunning, elaborately designed chambers, which exhibit the grandeur of elegant accoutrements. Consider the experience of entering a magnificent, yet strange mansion. In the owner's absence, one may strive to ascertain aspects of the owner's personality through a visual assessment of the architectural design, decor, and furnishings. The excitement of recognizing within each artifact a reflection of the taste and possibly personality of the owner becomes an intriguing task in itself.

So is it with God. As the Supreme Creative Intellect, the Divine *Persona* may never be known; and it therefore appears that a direct confrontation is impossible; nevertheless, we continue our search, and we are delighted when we discern a fresh facet of the Divine Personality emerging from our scrupulous analysis of definitive clues in nature.

Ultimately, we find that not only is a normative structure visible in the universe, but a unified communion of matter becomes more logically evident. A single strand of causal relationships appears to flow from the atom to the far-flung galaxies—in the refrain of the learned astronomer: " . . . the existence even of stars and galaxies depends in a delicate manner on the force of attraction between two protons"—and the radiant glow of the Divine Personality emerges. To such a Personality, we may find it possible to relate.

An ensuing dialogue between two personalities—one mortal—and One Divine—becomes not alone a possibility, but moreso, a human responsibility; and again, the strain of the learned astronomer resounds: "A remarkable and intimate relationship between man, the fundamental constants of nature and the initial moments of space and time seems to be an inescapable condition of our existence."

From this "remarkable and intimate relationship," there arises a discernment of moral law and order. It is the consequence of an exercise of extraordinary cognition between the I/mortal and the I/Divine. Man communicates with God, and majestically rises to the Prophetic Summit, from *whence* he may carry away divine tables of testimony inscribed with a meaningful ethos; as the lyrical prose of the learned astronomer avers: "Human existence is itself entwined with the primeval state of the universe and the pursuit of understanding is a transcendent value in man's life and purpose."

As man regards the grandiose panorama of his universe, and struggles to fathom its meaning, he is struck by the innate apperception that one of the great miracles of man is his ability to comprehend. Physiologically, the human rational organ consists of a complicated myriad of nerves, tissues, and brain cells. It has no rival on this planet in its capability for conscious awareness. Man is so greatly impressed with his own thinking mechanism that it appears correct for him to assume that his cognitive apparatus is a

169

keystone target of the universal design. Through this faculty of prodigious perception, species *Homo sapiens* is able to postulate meaning for existent matter and life.

Under the spell of sentimental romanticism, literary artists have assumed that God created an impressive universe for an awe-inspiring apprehension by man. They ponder: To what avail is the glory of creation and the imposing majesty of nature, if there be no other aesthetic consciousness to appreciate its splendor besides God? Empty, cold, and foreboding is the vast spectrum of starry space, if no conscious mind other than God perceives its ineffable beauty. Furthermore, man's appraisal of nature yields significant principles for conduct. It will be a tragic moment in the universe when human consciousness disappears, for then aesthetic judgment subsides, moral apperception fades into oblivion, and the comprehensible schema radiated from nature is lost to mortal awareness.

The learned astronomer taught that "the pursuit of understanding is a transcendent value in man's life and purpose." The measure of man and his status in relation to the universe and the Divine Being depends upon the manner of *understanding* that he seeks.

Biological Man achieves existential awareness of the crudest sort. He knows that things exist. Technological Man gains existential consciousness. He comprehends "how" things exist. Philosophical Man cognizes abstract awareness. He acknowledges a "why" in the universe. Theological Man attains moral perception. He searches for the "who" beyond the universe. Prophetic Man conceives moral discernment, for he discovers a Divine omniscient deontology operating behind the facade of the natural universe. He discerns the ultimate "Who," the Divine "I."

Finite man eventually comprehends the futility of seeking knowledge beyond the limits of human cognition. Although man becomes resigned to the hopeless situation of "never

170

to know," nonetheless, he is ever impelled to set forth on the quest for the ultimate source of all knowledge time and time again.

It is important that a perversion of the distinctive levels should not occur as a consequence of careless behavior. Such an event betokens a pathetic absurdity within the ranks of human existence. When Philosophical Man cavorts in the manner of Brute Biological Man, he distorts the human image and may be chastised for succumbing to idiotic behavior. When Domesticated Biological Man presumes to determine policy and practice for Theological Man, he may be castigated for appropriating unto himself an arrogance of license.

In a logical situation, from both a personal and communal standpoint, the higher plateaus should inspire and direct the lower categories so that they may endorse a humane and beneficial subsistence. However, when a superior sphere is subjugated and controlled by a lower level, a travesty of rational behavior may result. Life, then, becomes meaningless, and a fearful absurdity of being overwhelms the human population. Man cringes under the burden of existential insecurity; his survival is imperiled; and all hope for the future subsides. An illogical hierarchy convulses the stability of human development. The dire consequences evoke a mood of deep anguish, leading to futility, despair, and inauthentic existence.

What is the key to authentic existence? How shall man overcome the ominous threat of anguish and despair? Man must determine that all is not futile and hopeless; that it is possible to "apprehend the ethos of the evening star"; and that this noble task may be accomplished by drawing ever closer to God.

Man must seek to relate to God in a personal intimate fraternity. If intimacy suggests a romantic affair with God —so be it. Indeed, man has the capacity to indulge in an

171

affair of intellectual romance with the Divine Personality, and he should not overtly evade this splendid opportunity. A romance-experience with God directs man to his finest hour of authentic existence. There are, however, certain prerequisites.

The Divine romance-experience cannot be fulfilled on the lower Biological Plain; nor can it be consummated on the Intellectual Plateau in the regions of Technological and Philosophical Man. Lower Biological Man sees only his own ego; with him, narcissism is possible, but not romance. Amour is futile on the Technological Plateau, because it is absurd to establish an affair of the heart with an "it." Philosophical Man is deficient, because one cannot enter into a reciprocal love affair with a "He."

The first meaningful opportunity for a Divine romance occurs on the Theological Plateau. A liaison becomes possible when God is envisaged in an "I–Thou" relationship. On the Prophetic Summit, where man exercises his maximum intellectual capacity by aspiring for an "I/mortal–I/Divine" encounter, the Divine romance-experience blossoms into its fullest expression.

Where should man seek his rendezvous with God? The atmosphere of the research laboratory is hardly conducive for such a demanding encounter; nor is the marketplace any better; nor is the mystic atmosphere of a guru parlor propitious for such a courtship. Where, then, indeed shall man turn? The answer is remarkably simple.

In view of the unique nature of the human aesthetic character, the most desirable appointment for a Divine meeting occurs in the House of Prayer. Man's ability to experience and express deep-felt emotions makes it appropriate, as well as practical, to identify with a respectable and a well-established theological discipline. The individual passion for Divine adoration is best suited to the milieu and format of formalized religion. A more meaningful religious

172

experience may be developed in a theologically dominated environment. Although a Divine communion may be consummated almost anywhere, God seems to be more readily accessible in a respected place of worship.

Plainly stated, the gateway to God is available to all. To reach it, one need only seek the Divine Thou as echoed by the fervent cry of His dedicated servant, the prophet Jeremiah. In a historic hour of deep anguish and despair, he implored with heartfelt simplicity: "Turn us unto Thee, O God, so that we may return."

Index

174

175

176

Timaeus (Plato), 98

Time: as a fourth dimension, 101; in relation to vacuum, 101; sidereal time, 98; solar time, 98; and space, 30, 31, 37, 98, 99; time order, 100; time/spatial coincidence, 145. *See also* Space

Transcendency, problems relating to, 100

Transcendental idealism. *See* Idealism

Uniformity in the universe, 80–81, 98–99. *See also* Isotropy

Universal, 144–45. *See also* Particular

Utilitarianism, 55

Voltaire, Francois-Marie Arouet de, 65

Ward, James, 55

White dwarf, 128. *See* Stellar collapse

"Who," 84, 118, 119, 123, 129, 130, 136, 137–39, 140–41, 145, 158, 167, 168, 170; deviated "Who," 140

"Why," 68–74, 84, 95, 107, 108, 110, 116, 117, 136–39, 140, 167, 170; absurd "why," 72; avoided by Technological Man, 116; cosmic "why," 72; normative relationship, 136, 141; personal "why," 72

Wilberforce, Bishop Samuel, 38

Wordsworth, William, 55

Wright, Thomas, 29

Zechariah (prophet), 14

"Zone of Life," 144